American Design

MoMA Design Series

American Design

Russell Flinchum

MoMA

Produced by the Department of Publications, The Museum of Modern Art, New York, and 5 Continents Editions, Milan

This publication was made possible by the Nancy Lee and Perry Bass Publication Endowment Fund.

Published by The Museum of Modern Art, 11 West 53 Street, New York, New York 10019

Distributed in the United States and Canada by D.A.P./Distributed Art Publishers, Inc., New York

Distributed outside the United States and Canada by Abrams Books, London

Library of Congress Control Number: 2008932783
ISBN: 978-0-87070-740-7

This book is typeset in Neue Helvetica by Duke & Company, Devon, Pennsylvania. The paper is R4 Matt Satin 170 grm^2

Color separation by Eurofotolit, Milan

Printed and bound by Conti Tipocolor, Florence, Italy

Designed by Dondina Associati

The Museum of Modern Art
Rebecca Roberts editor

5 Continents Editions
Debbie Bibo editorial director
Laura Maggioni editorial coordinator
Annarita De Sanctis layout

Credits

Photograph p. 97 © 2008 Estate of Harry Bertoia/Artists Rights Society (ARS), New York; p. 66 © 2008 The Isamu Noguchi Foundation and Garden Museum, New York/Artists Rights Society (ARS), New York; p. 150 courtesy Target Corporation; pp. 55 and 71 © 2008 Frank Lloyd Wright Foundation/Artists Rights Society (ARS), New York

Photographs by Phil Banko: p. 149; Department of Imaging Services, The Museum of Modern Art: pp. 16, 19, 20, 23, 24, 27, 34, 36 (both), 38 (both), 41, 42, 49–54, 56, 59, 60–65, 67–69, 71, 74–77, 82, 84–93, 95, 96, 98, 99, 102–104, 108, 110–14, 117–23, 125–28, 130, 132, 134, 135, 138, 139, 140–42, 144–48, Jon Cross/Erica Staton: pp. 58, 66, 116, 124, and 133, Thomas Griesel: pp. 48, 94, 115, 151, S. Joel: pp. 14, 57, 80, 97, Kate Keller: pp. 109, 137, Jonathan Muzikar: pp. 37, 55, 70, 78, 81, 83, 100, 101, 105, 143, Mali Olatunji: p. 100

Images pages 153–55: Brooklyn Bridge, c. 1915. Courtesy Library of Congress Prints and Photographs Division, Washington, D.C.; Golden Door, World's Columbian Exhibition, Chicago, 1883. Courtesy Chicago History Museum, ICHi-17384; Kodak Brownie Camera. Courtesy Strong National Museum of Play®, Rochester, N.Y.; Ford Model T, 1916. Courtesy Science and Society Picture Library/The Image Works; Harley Earl and Lawrence Fisher in a General Motors LaSalle Convertible, 1927. Courtesy Harley Earl Inc./Richard Earl; Franklin Delano Roosevelt, c. 1933. Photograph by Elias Goldensky. Courtesy Library of Congress Prints and Photographs Division, Washington, D.C.; four P-51 Mustangs flying in formation, 1945. Photograph by Toni Frissell. Courtesy Library of Congress Prints and Photographs Division, Washington, D.C.; Julius and Ethel Rosenberg, 1951. Photograph by Roger Higgins. Courtesy Library of Congress Prints and Photographs Division, Washington, D.C.; Malcolm X, 1964. Photograph by Marion S. Trikosko. Courtesy Library of Congress Prints and Photographs Division, Washington, D.C.; cover of the record album *More Songs about Buildings and Food,* by the Talking Heads, 1978. Under license from Warner Bros. Records; President Ronald Reagan and Nancy Reagan in inauguration parade, 1981. Courtesy Ronald Reagan Library; Hubble Space Telescope, 1997. Courtesy NASA; Apple iPod, 2001. Photograph by Thomas Griesel, Department of Imaging Services, The Museum of Modern Art

In reproducing the images contained in this publication, the publishers obtained the permission of the rights holders whenever possible. If the publishers could not locate the rights holders, notwithstanding good-faith efforts, they request that any contact information concerning such rights holders be forwarded so that they may be contacted for future editions.

Paola Antonelli

The Land of Plenty

American design, like much of American culture, perennially oscillates between populism and elitism, between the revolutionary beauty and availability of Tupperware and the elusive exclusivity of Tiffany. Both extremes express design excellence, and since its inception in 1929 The Museum of Modern Art has documented its country's material culture with a selection of the best examples from the whole spectrum of furniture, product, and graphic design. The Museum's very first acquisitions in design were a group of more than one hundred American industrial objects—propeller blades, coils and springs, laboratory appliances, tools, household objects, and furniture—that had been displayed in *Machine Art,* the 1934 exhibition of design that architect Philip Johnson organized with the assistance of the Museum's founding director, Alfred H. Barr, Jr. At the same time, the curators were also examining the sophisticated modernism of the protagonists of the Bauhaus school. This attention to popular and utilitarian design and to the high end of design research is still very much in evidence today in the Museum's exhibitions, programs, and acquisitions.

Although the United States was founded in opposition to the class systems of Europe, at the beginning of the nineteenth century America was still psychologically a European colony. American high society strove to emulate the refinement of the European aristocracy; thus, furniture for the wealthy was generally imported, and the middle class, which longed for the same level of luxury and style, bought imitations. The first truly indigenous American modern design, excluding practical adaptations of European vernacular objects such as the American tree-feller's ax and isolated cases like Samuel Gragg's bentwood chair of 1808, was created by the

Shakers, a religious group in the country's northeast. Shaker furniture and interior architecture displays sobriety and honesty in a direct portrayal of the circumstances that generated them. Available materials were used in harmony with their capabilities, according to what American design historian Arthur Pulos called "the principle of beauty as the natural by-product of functional refinement"—a principle that for many designers in the twentieth century summed up design perfection.

The Industrial Revolution had raised the specter of a world in which visible differences among classes could be erased by mass-produced objects available to all and functional wealth would be evenly distributed. In the United States, Henry Ford took a step toward the realization of this dream with his Model T experiment of the 1910s, and later in the century, in the 1950s, abundance and opportunity advanced apace. Especially on the west coast and in the midwest, the resources of the idle war industry were converted into civilian and domestic use by great designers like Charles and Ray Eames and by anonymous engineers, providing the booming middle class with a brand-new, clean, efficient, and (most importantly) affordable world. The east coast architectural aristocracy, led by The Museum of Modern Art and Harvard University, embraced the radical and rarefied visions of illustrious modernist émigrés Walter Gropius and Ludwig Mies van der Rohe, but the new architecture and design "for modern living" coming from the west and the midwest was the flagbearer of American sensibility and idealism. Born out of economy and equality, it was about practice, not theory.

In the United States, consumer products are status symbols, and the advertising companies clustered on Madison Avenue in New York were responsible for shaping the dream of social mobility through consumption. Since the 1930s, they have found powerful allies in design masters such as Walter Dorwin Teague, Norman Bel Geddes, and, last but not least, Raymond Loewy. Loewy, who is also credited with inventing the profession of designer, added a mythic quality to products such as the 1953 Studebaker Starliner Coupe by endowing them with powerful streamline accents and brand mystique, transforming them into objects of desire.

Designers like the Eameses, Harry Bertoia, Eero Saarinen, and George Nelson, well represented in MoMA's collection, found their creative interlocutors in the 1940s in a handful of visionary, family-based companies that included Knoll and Herman Miller. Businesses such as these, with low overheads, are best able to support innovation and take risks—until they grow into multinational corporations, which cannot afford the same agility. The inadequate number of midsize manufacturers in the United States today has stifled the formation of a strong group of versatile designers of the kind enjoyed by most European countries. Until recently, most talented American designers were forced to manufacture their products in craft workshops at high cost and in limited series, although most of them would have liked to counter the notion that good design is a costly added value.

Starting in the late 1990s, a new generation of media savvy, internationally trained American designers has established productive relationships with manufacturers in other countries, and mass production has made a comeback. Some manufacturers are attempting to incorporate high-end brands and signatures into the products, melding populism and exclusivity. Several large American retail companies have signed on celebrity designers to market "quality" products at affordable prices, in order to counteract the increasing appeal for the American market of such exotic chains as Ikea and Conran. For example, the Minnesota-based department-store chain Target has produced a range of more than two hundred products for the home designed by American architect Michael Graves. The Good Grips line of home utensils and the 3M Post-it Note, both based on scientific research (and both in the Museum's collection), are pervasive and mass-produced milestones of design history. And then there is the iPod, which the Museum acquired in 2001, the year of its release. The extent to which this now ubiquitous cultural artifact revolutionized not only the genre of portable music devices but also the expectations of buyers and users worldwide cannot be underestimated. It elevated all standards and gave the world renewed respect for American design.

Once again, Americans are succeeding at the game they play better than anyone else: not only image and brand design in the land of plenty, but great, solid popular design. In keeping with MoMA's mission, which in the words of Alfred Barr supports "the study of modern arts and the application of such arts to manufacture and practical life," the Museum joins the celebration with enthusiasm and a critical spirit.

Russell Flinchum

Vitality and Ingenuity

Historians of modern design have often invoked 1851 as the starting point of modern design consciousness, with Joseph Paxton's Crystal Palace in Hyde Park, London, containing the "Works of Art and Industry of All Nations," as the key monument—a building whose radical nature generally outstripped its contents in terms of its modernity of appearance and construction. Adrian Forty pushed that date back to the eighteenth century and the work of Josiah Wedgwood in his well-received 1986 study *Objects of Desire: Design and Society Since 1750.* Christopher Dresser is sometimes referred to as "the first independent industrial designer,"[1] and his corpus resides pretty firmly in the last quarter of the nineteenth century. An even stronger claim has been made for Peter Behrens and his work for the AEG (the German electric company) beginning in 1907.[2] None of this is very helpful in trying to determine when design "began" in the United States. I postulate that it started with the publication of *The Young Mill-Wright and Miller's Guide,* by Oliver Evans, in Philadelphia in 1795. While Evans is rarely discussed outside history-of-technology circles, his contributions to the Industrial Revolution in America were so seminal—the high-pressure steam engine being his most notable invention—that he outshines Robert Fulton (whose steamboat used low-pressure steam, the technology of James Watt) and Eli Whitney (whose star has dimmed considerably as "the father of mass production in America," since he did not, in fact, manage to mass produce anything, in spite of his claims).

Why Evans? Because Evans did not merely foreshadow but realized in his work America's most important contribution to modern design: *flow.* His book unveiled a system, utilizing many of his own inventions, that automated the manufacture of flour from grain in a continuous process that called for no human labor or intervention when properly set up. There are no objects designed by Oliver Evans in this book. Yet flow was the final component successfully added to the mix that made "art and industry" into modern American design as we know it; it was the critical element in realizing a true system of mass production out of the many manufacturing advances made over the hundred years after Evans. Henry Ford's first mass-produced Model T rolled off the assembly line in Highland Park in 1913. Mr. Ford's car demonstrated to an international audience that products near the cutting edge of technology could be provided in huge numbers at a reasonable cost. It revolutionized the world. It is a world we live in today. Consider for a moment how many of the items within your view are unique. Now consider how many are identical to hundreds of thousands or millions of others. Now consider that your personal sense of style is determined largely by your choices within the latter, with a few of the former tossed in for effect.

The curious environment that was the early United States can be seen in two objects in the collection of The Museum of Modern Art: an ax (c. 1946) and a bentwood chair (c. 1808). There is no designer of the American ax and no specific date at which it was designed. It morphed from seventeenth-century examples brought from England, well suited to use by skilled woodworkers for a number of tasks, into a very different item that is best suited for felling trees. Along the way, its cutting edge reoriented itself ninety degrees and its handle doubled in length (and double-faced versions doubled the durability of its edge). This sort of anonymous design has long posed problems for design historians, and it provides one of the underpinnings or truisms of much of design writing: Design can evolve without designers. This mythology was cemented through works by two of the better theorists of the modern movement, Le Corbusier's *L'Art décoratif d'aujourd'hui* (*The Decorative Art of Today*) of 1925 and Siegfried Giedion's *Mechanization Takes Command: A Contribution to Anonymous History* of 1948, both of which were at least partially inspired by the author's exposure to American products. In both men's writings, it sometimes seems that designs are "better" when they evolve devoid of specifics, as if moved by greater societal forces.

And then there is the ax's antithesis, Samuel Gragg's Side Chair, the creation of an individual at a specific time and place. It did not seem to merit much attention in design circles until recently, in spite of demonstrating a perfected system of bentwood construction a good forty years before Michael Thonet and others began making chairs through similar methods. Gragg's chair, in spite of its appealing simplicity and revolutionary construction, did not displace its contemporaries; historians have viewed it as a curiosity, a protodesign, perhaps, but not part of the larger narrative. We can't be sure of the insights that led to Gragg's design, although boatbuilding seems a likely source of inspiration and quite plausible given the location of his workshop, near the Boston waterfront. We're not even very well equipped to look at the chair properly: Most of us would be satisfied with the description of its materials as "wood," but to furniture-makers in early-nineteenth-century America, the key question would be not what kind of wood, but what kinds of wood! They would have been no more content with a single-word description than we would be with a label for Marcel Breuer's Wassily Chair that described its materials as "stuff." And then there is the real theoretical problem: The chair is decorated, and to Gragg, who does not seem to have executed its decoration but hired someone else to paint it, that decoration was as integral a part of the final product as the elegance of its line or its ability to support weight. But, as Le Corbusier assured us in 1925, "the decorative arts of today are undecorated." Was Gragg's chair passed over for consideration by earlier generations of design historians because it was painted with decorative patterns? Or were they simply unaware of it?

This brief discussion of only two objects gives us an idea of just how difficult it is to establish the parameters of design. In this book, exclusions are the norm rather than the exception: There are few works of craft and no works by European émigrés who continued to practice in their accustomed vein after their arrival in the United States—and those are just the author's exclusions. Other than an admirable lacquer screen by Eileen Gray, there are no designs that can be grouped under the rubric Art Deco in MoMA's collection, nor are there many examples of American streamlined design from the 1930s and 1940s (Egmont Arens and Theodore C. Brookhart's Streamliner Meat Slicer, c. 1940, is one of the exceptions, and, like Gragg's chair, it is a latecomer to the collection).

Eileen Gray
(British, born Ireland.
1879–1976).
Screen, 1922.
Lacquered wood
and metal rods,
6′2½″ x 53½″ x ¾″
(189.2 x 135.9 x 1.9 cm).
Manufacturer: Eileen Gray
Workshop, Paris, France.
The Museum of Modern
Art, New York.
Hector Guimard Fund,
1978

Because of its brevity, this book cannot serve as a comprehensive introduction to design in the United States; and, as all of the examples are drawn from MoMA's holdings, it is a selection from a selection. It's not incorrect to say that the design component of the Museum's Department of Architecture and Design has had an attitude from the start. The objects in the collection are a reflection of the very pronounced tastes of a handful of individuals, which were hardly uniform in spite of their many commonalities. While often scholarly, the directors, curators, and staff who assembled the collection were activists rather than academics. And while their mission has undergone distinct eras, each with its own approach, the goal has remained consistent: to put the best of modern design before the public, with the aim of influencing members of that public's choices in equipping their own lives. That influence has varied from the didactic and educational to approaches that came close to branding MoMA's own choices in the marketplace. Much heat (and some light) has been generated around the discourse that evolved between the Museum's curators and their critics; let us stop for a moment to consider the remarkable achievement of shaping a coherent collection over a period of almost seventy-five years, a period that has seen an incredible deluge of design reaching the marketplace, ranging from the timeless to the flavor of the moment. Love it or hate it, MoMA's design collection has been the starting point for most discussions of what is modern about modern design.

It is hoped that the reader will be struck by visual commonalities and themes among the objects here and perhaps even be a little disturbed by them: Why show two bottle openers, two pairs of scissors, two teakettles? Because this book is about design in the United States, and Americans believe in choice. In an effort to reach some understanding about what is American about American design, it is possible to say quite clearly what it is not: it is generally not about formal purity and the refinement of a design into an ideal. For a country obsessed at the turn of the last century with the "one best way" to do a task, as envisioned by Frederick Winslow Taylor, the time and motion efficiency expert, Americans have been quite undoctrinaire in their approaches to similar challenges, designing bottle openers—such as Harvey J. Finison's (1977) and Henry Altchek's (1980)—that have nothing in common visually, in the way they are made, or even in the way they are handled, yet work perfectly well, and scissors that apply a new material in different manners—those by Clair H. Gingher (1979) and Alan Spigelman (1980), for example. A "boil-off" between Lurelle Guild's and John

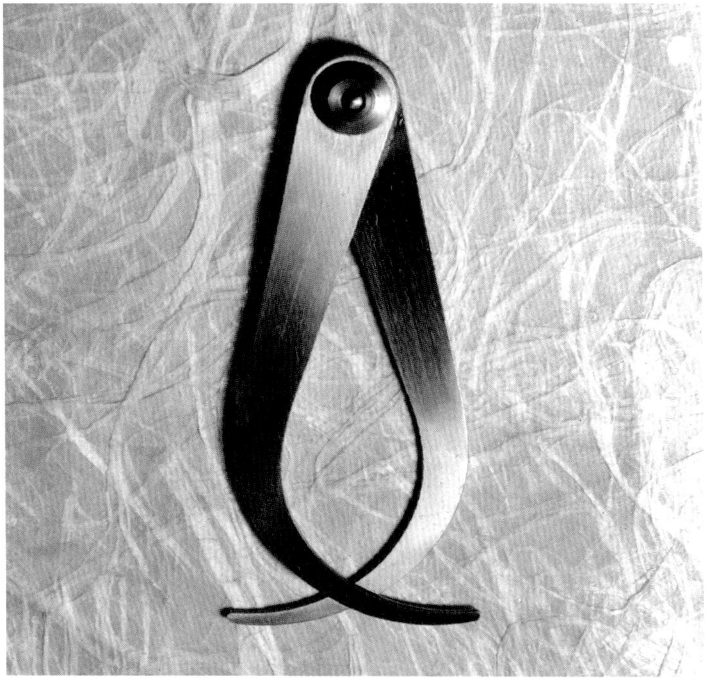

Brown & Sharpe Manufacturing Company (USA, established 1833). Outside Firm-Joint Calipers. Before 1934. Tempered steel, 8½ x 6½ x ½" (21.6 x 16.5 x 1.3 cm). Manufacturer: Brown & Sharpe Mfg. Co., Providence, R.I. The Museum of Modern Art, New York. Gift of Brown & Sharpe, 1934; photograph by Ruth Bernhard

G. Rideout's teakettles might not establish the best; each has its merits and its drawbacks, and neither has emerged as what Le Corbusier characterized as the *objet type* in the same way that the wooden pencil, the paperclip, and even the Band-Aid constitute the thing itself. The classic statement of high modernism may be "The solution is to be found within the correct formulation of the question," but the American response has generally been to provide a couple of solutions to the original question then move on. Perhaps the frontier mindset lingers, although the frontier had disappeared long before most of these objects were created.

American design, for better or worse, has largely been about providing pragmatic, short-term solutions to problems, unlike the European emphasis on tradition. Part of this lies with our national restlessness and our usually optimistic view that the future will provide better solutions through the advent of new materials, processes, and expertise. The downside has been a sometimes reckless disregard for our environment, a lack of concern for historic preservation, and an emphasis on disposability and the annual model change. The upside has been an explosion of design largely unfettered by tradition, dissatisfaction with received wisdom and notions of how things should be done, and a healthy belief that just because something is popular doesn't mean it is suspect. American culture is, by and large, popular culture. If it appears at times a little vulgar, a little shoddy, derivative, or banal, don't worry —something else will come along any minute.

The development of manufacturing technology in the United States was fundamentally different than what unfolded in Europe. American talent found itself less bound to the cities and to existing networks of professional expertise and was freer to travel to where opportunities abounded. One of the few places it pooled was in the federal armories tasked with making firearms, and it was here that the fundamentals of "armory practice," with its prioritization of the manufacture of identical and interchangeable components, were established.[3] Those practices were disseminated by workers who left the armories for other opportunities, and in the course of the nineteenth century these skilled workers became known as machinists. Machinists were rarely specialists, but they brought with them to a new workplace a method that emphasized the use of a rational system of jigs and fixtures (to identically align components being worked on) and standards and gauges (to accurately measure their results). The Outside Firm-Joint Calipers manufactured by Brown & Sharpe are typical of the tools they used. Wherever machinery could be used to substitute for human labor or to make use of unskilled or semiskilled hands, it was generally implemented, even if it was initially costly. From firearms and clocks to sewing machines and bicycles, these undertakings grew more complex, and the knowledge base grew accordingly. Without the processes (such as electrical resistance welding) and materials (high-strength alloyed steel) developed in the United States to manufacture the bicycle, the automobile probably would have remained a luxury item here until well after World War I, as it did in Europe. Architect Frank Lloyd Wright's client Frederick C. Robie of Chicago was a manufacturer of top-quality bicycles. Walter Chrysler of Chrysler Motors was a machinist who displayed his tool kit with pride in the Cloud Club of his eponymous building in New York (until the Empire State Building exceeded it in height a second time, surely one of the sadder footnotes in his biography).

Having a pool of highly skilled and flexible manufacturing mercenaries did not guarantee beautiful products, however. The "engineer's aesthetic" so prized by the first generation of modern architects in Europe might have been inspiring on the lofty level of suspension bridges and skyscrapers, but it left a great deal to be desired when it came to home appliances and the other equipment of twentieth-century domesticity. If engineers could "do for a dollar what any fool could do for two," they generally did not concern themselves with the fripperies and gewgaws associated with making new conveniences pleasing to consumers, and thus the modernist critiques of "applied decoration" were generally right on the money. Aesthetics were often an afterthought.[4] The enclosed, all-metal-construction American automobiles of the early 1920s often resembled nothing so much as "telephone booths on wheels," as one contemporary described them, and early toasters, electric stoves, and bathroom fixtures fared little better. Parisian-born designer Raymond Loewy was shocked upon his arrival in the United States from Paris to find that the homeland of telephony, mass production, electric traction, and heavier-than-air flight was an ugly, "supervolted" horror show.[5] There was a great deal of room for improvement.

That is not to say that Secretary of Commerce Herbert Hoover was entirely correct when he declined on behalf of the nation the invitation to participate in what would become known as the "Art Deco" International Exhibition of 1925, in Paris, writing that the United States has "no original works to show."[6] Original modern works had been around for a while. One look at Will Bradley's cover for the Thanksgiving 1895 edition of the literary magazine *The Chap Book* reveals a full command of the Art Nouveau aesthetic, echoing in its abstraction Henry Clemens Van de Velde's well-known poster for Tropon—a salad oil, by the way—that is also part of the Museum's collection of graphic design. Frank Lloyd Wright had written vigorously about the role of the machine in design in *The Ladies' Home Journal* in 1910; Gustav Stickley's Armchair (c. 1907) in the Mission Style embraces the aesthetic of the crosscut saw and repetitive vertical forms. Wright's Office Armchair (1904–06), designed for the executive offices of the Larkin Administration Building in Buffalo (completed in 1905 and demolished in 1930), was a microcosmic statement of an all-encompassing vision of modernity and a revolutionary rethinking of the American workplace. Its gridded, ventilated back support echoes the contemporaneous aesthetic of the Wiener Werkstätte, and it was also mobile, unlike the seats of the building's many office workers, who labored at desks among

Henry Clemens Van de Velde (Belgian, 1863–1957). *Tropon, l'aliment le plus concentré*. 1899. Lithograph, 44 x 30⅜" (111.8 x 77.2 cm). Printer: Hollerbaum & Schmidt, Berlin. The Museum of Modern Art, New York. Arthur Drexler Fund, 1988

built-in filing drawers around a central atrium with inspirational mottoes carved into its upper surfaces. It is no wonder that Wright's *Ausgeführte Bauten und Entwürfe* (*Studies and Executed Buildings*) hit Europe's architecture establishment like a bombshell when it was published in Berlin in 1910, just as three young architects were passing through the office of Herr Dr. Professor Peter Behrens: Walter Gropius, Ludwig Mies, and Charles-Édouard Jeanneret.[7]

Yet it's a good idea not to overstate the case for America's design sensibility before the late 1920s. From the advent of the Armory Show in 1913 until the late 1920s, it was an age of artisans in industry rather than designers shaping products for mass production. The Corona Typewriter (1912) is a very good example of such design for that year; it actually approaches some unity of appearance, unlike many of its contemporaries. Charles Dana Gibson's *"Here he is, Sir"* recruiting poster for the United States Navy (1917) seems almost faux-naive coming from the sophisticated hand of the creator of the Gibson Girl—but the young recruit was modeled on the artist's son Lang, and the earnest quality of the image likely reflects Gibson's own sense of sacrifice.[8] Coors Porcelain Company's distinctly modern Acid Pitcher (c. 1930) was copied directly from a German model from before the outbreak of World War I, when the British blockade ended the export of laboratory ceramics and glassware (and, interestingly, just at the juncture when Corning Glass Works of New York was perfecting Pyrex, a borosilicate glass that was ideal for both labware and oven-to-table dishes). The United States was woefully unprepared for its entry onto the European stage in World War I, and the nation was about to grow up real quick. Warfare on the Western Front was the end of America's idealism and

Russel Wright
(American, 1904–1976).
Armchair with Adjustable
Back. 1932. Mahogany,
leather, and pony fur,
31½ x 29 x 27″
(80 x 73.7 x 68.6 cm).
The Museum of Modern
Art, New York.
Purchase Fund, 1958

the forge of an entire generation of leaders who would come to the fore in the next war: Dwight Eisenhower, Douglas MacArthur, “Wild Bill” Donovan, and an obscure lieutenant from Missouri with the middle name S who would succeed Franklin Roosevelt as president of the United States in 1945.

Modern design did not originate in the United States, but the practice of industrial design was perfected here, largely in New York between 1927 and the outbreak of World War II. American exposure to the 1925 International Exhibition in Paris was the driving force behind the revolution in aesthetics that would begin just a few years later. Designers visiting Paris that year carried away impressions of style and sophistication and upon their return found ways to approximate these for domestic consumption. They were greatly aided by European émigrés, many of whom had received academic instruction at the highest levels. Austrian Paul Frankl's objectives for modern design went far beyond his Skyscraper bookcases, and Walter Von Nessen, who had studied under Bruno Paul in Berlin prior to World War I, began a quiet revolution in domestic lighting with his seminal swing-arm Desk Lamp (1927). American native Russel Wright joined Von Nessen in designing dozens of housewares for Chase Brass & Copper Company, including a Pitcher (c. 1921–41), sometimes using preexisting components from the company's plumbing line as the kit of parts for decorative items such as bookends. It was just about this time that The Museum of Modern Art burst upon the scene.

The famous *Machine Art* exhibition of 1934, organized by Philip Johnson, was not the first design show held at the Museum (a modest show he assembled in 1933, *Objects: 1900 and Today,* holds that distinction), but to contemporary critics it seemed like a one-two punch to popular taste following 1932's *Modern Architecture: International Exhibition.*[9] Working within a shoestring budget and largely with items lent or donated by manufacturers, Johnson managed to be both high-minded (including quotations from Plato's *Philebus* in the show's catalogue) and provocative. It's worth noting that MoMA's exhibition was only one of three major shows on design in Manhattan that year. The Metropolitan Museum of Art featured modern interiors and furnishings in its exhibition *Contemporary American Industrial Art,* which stressed collaborations between architects and industrial designers, and the short-lived National Alliance of Art and Industry displayed mimeograph machines, refrigerators, and a number of other highly complex and highly styled items at Rockefeller Center. But

Machine Art was the exhibition that caught the popular (or, at least, the critical) imagination. Massive springs, propellers, and a certain self-aligning ball bearing first seen in that show are still encountered by today's visitors—even as MoMA's definition of modern design has grown increasingly complex. Even then, it was more inclusive than one might expect. It embraced Lurelle Guild's Wear-Ever Teakettle (c. 1923–33), just one of dozens of items he designed for The Aluminum Cooking Utensil Co., New York, including an entire line dubbed Kensingtonware; Guild was not shy about using historicizing elements in his pared-down aesthetic. Like Russel Wright and Von Nessen, he, too, designed for Chase. And one has to wonder if he might have had a hand in another design MoMA exhibited, the Wear-Ever Rotary Food Press (1932); its support is just a little too elegant next to its competitors, no matter how generic it first appears. John G. Rideout may have operated out of Cleveland, Ohio, not New York, but his Teakettle (1936), also shown at MoMA, is just as typical of the 1930s as Guild's. If not streamlined, it is, in the parlance of American design pioneer Henry Dreyfuss, "cleanlined," with a dash of "survival form"[10] (its lid is semipermanently attached and may be removed only by first detaching the kettle's handle with a screwdriver). Introducing Magnalite (a new alloy of magnesium, aluminum, and nickel), it epitomizes an American approach to design during a period in which the country struggled to pull itself out of the Great Depression.

When Johnson left the Museum (temporarily), John McAndrew took over duties in the Department of Architecture and Industrial Design and organized *Useful Household Objects Under $5* in 1938. This inaugurated one of the Museum's growing concerns—demonstrating that good design was not only available to the public, but it wasn't necessarily costly either. When a young architect named Eliot Noyes took over from McAndrew, fresh from working with Walter Gropius and Marcel Breuer, he brought both a skilled eye and a distinctly, well, WASP-ish view of the importance of integrity in design that had meshed well with his European mentors.[11] The exhibitions he assembled as director of the Department of Industrial Design (distinct from the Department of Architecture between 1940 and 1949) were far less doctrinaire than one might at first expect. There was room for "Americanness" in Noyes's conception of design, such as that exemplified by Richard Kelly's simple Table Lamp (c. 1940) that just made the price cutoff in the 1940 exhibition *Useful Objects of American Design Under $10.*[12] Noyes soon joined the war effort, but he would be back, both as an

Walter Dorwin Teague (American, 1883–1960). Vase. 1932. Glass, 10⅛ x 11" (25.7 x 28 cm). Manufacturer: Corning Glass Works, Steuben Division, Corning, N.Y. The Museum of Modern Art, New York. Gift of the manufacturer, 1934

Gilbert Rohde
(American, 1894–1944).
Chair. c. 1938.
Stainless steel
and Plexiglas,
31½ x 17½ x 21"
(80 x 44.4 x 53.3 cm).
The Museum of Modern
Art, New York.
Gift of the Gansevoort
Gallery, Jeffrey P. Klein
Purchase Fund,
and John C. Waddell
Purchase Fund, 2000

organizer and as a designer in his own right. Edgar Kaufmann, Jr. (whose businessman father had commissioned a country retreat from Frank Lloyd Wright that would be dubbed Fallingwater) was also on the staff of the Department of Industrial Design. In spite of the fact that he was an aesthete educated in Vienna, Kaufmann's taste had a remarkable coincidence with that of Noyes. The two men could not have been more different in background and temperament, and this common ground is perhaps indicative of the power that the words "good design" held at the time.

As the 1930s progressed, the aesthetic of the Machine Age evolved under the Museum's aegis into that of Organic Design, "the principle of organic integration of function, technology, and form," as the Museum defined it, which found an iconic expression in Russel Wright's "American Modern" line of stoneware for the Steubenville Pottery Company (1937). Wright and his wife, Mary, were a powerful marketing team; she was the first to use "blond" to describe the light-colored, less costly woods they sought to introduce in furniture. "American Modern" came in colors like Seafoam Green and Bean Brown, suggesting the more lighthearted approach to home entertaining they advocated. The Radio Nurse Speaker (1937), a rare industrial design by Isamu Noguchi, brought considerable sculptural qualities to the handling of Bakelite. Meant to protect toddlers electronically in the wake of the Lindbergh kidnapping, the device layered the comforting outline of the nurse's headdress common at the time with the more aggressive evocation of a fencer's mask. (The sheet-metal Listening Ear that sat next to the child's crib didn't seem to receive any design attention at all. Then again, toddlers weren't doing the buying.) The design of the Ironrite Ironer Company's Health Chair (1938) makes sense when we realize that an Ironrite Ironer was operated from a seated position (it was a "mangle," in British parlance, but the manufacturers did everything to avoid that term, for obvious reasons). One can only speculate whether Ironrite owner Herman Sperlich and his company's engineers had seen Breuer's very similar chairs manufactured in Zurich; by the end of the decade, publications showing images of modern furniture designs were commonplace, and Americans were growing increasingly sophisticated—and they had never been particularly shy about copying a good idea without attribution, one reason why the United States enjoys a patent system that dates back to the founding of the nation. Collaborative designs by Charles Eames and Eero Saarinen (chairs, benches, and cabinets, among others) drew much attention when they debuted in 1941 in the exhibition *Organic*

Design in Home Furnishings at MoMA, with the chairs' complex plywood shells bent in three planes, albeit shown covered with upholstery; another trend is evident in Carl Anderson and Ross Bellah's rattan Sectional Chair (c. 1940). A traditional material was used in an innovative manner—notice the cross-bracing of the seat and the fact that the rattan is used in large, uniform circumferences—as if the Machine Age's metal tubing had been touched with the Organic Fairy's wand. When it was first displayed, the fabric covering its cotton upholstery pads bore a lively surface design by Noémi Raymond, wife of the Czech-born American architect Antonin Raymond. Perhaps the collaborative nature of most of the pieces on view in *Organic Design* mirrored a growing consciousness that the next effort—beating back the Axis—was going to demand a lot of collaboration of a positive kind.

Glenn Grohe's wartime poster *He's Watching You* (1942) must have done an admirable job of keeping workers on their toes. The outline of the Wehrmacht helmet coupled with the slit, reptilian eyes of this lurking figure, who approximates a stealthy submarine, is surely one of the more chilling reminders of just how dire the situation appeared from the end of 1941 well into 1943. Americans (and MoMA) leapt into the war effort, and on the home front that meant rationing. The Museum's 1942 exhibition *Useful Objects in Wartime Under $10* was accompanied by a brief publication with images of typical consumer items nixed with bold brushstrokes, a device lifted directly from the poster for *Die Wohnung,* the Deutsche Werkbund's exhibition of interiors that accompanied the Weissenhofsiedlung showcase of modern architecture in Stuttgart in 1927. Steel, chromium, tin, aluminum, Lucite, Plexiglas, Bakelite, and Beetleware (another early plastic) were all *verboten* because of rationing.[13] Glass and ceramics were advocated; Corning Glass Works' Double Boiler, first seen in 1938 in *Useful Household Objects Under $5,* made another appearance at the Museum. When the catalogue for the exhibition *Art in Progress,* a survey rather than a celebration of the first fifteen years of The Museum of Modern Art, was published in 1944, the section on design was entitled Design for Use. Organic design was the watchword of the day, seen in furniture like Jens Risom's Low Lounge Chair (1941). Another surprising example of seating design is the Gun Turret Seat (c. 1939–44) designed by the engineering staff of McDonnell Aircraft Corporation for the B-24 bomber using an innovative combination of paper and plastic dubbed "Structomold"—developed for lightness and strength, one hopes, not disposability.

George Nelson
(American, 1908–1986).
Mobile Table
(model 5152). c. 1949.
Angle-iron frame
and plastic,
24¾ x 24½ x 17¼"
(62.9 x 62.2 x 43.8 cm).
Manufacturer: Herman
Miller Furniture Co.,
Zeeland, Mich.
The Museum of Modern
Art, New York. Gift of the
manufacturer, 1953

Perhaps because American designers and their creations so dominated the worlds of architecture, fine art, and design following World War II, we all carry around some internalized notion of what constitutes the Americanness of their creations. Modernism, of course, was always meant to be an escape from the parade of styles that characterized the nineteenth century, almost all of which had their American versions and interpretations. At its loftiest peak, modernism seemed to promise a perpetually renewed built domain based on innovations in materials and processes that would each rationally overturn its predecessor. This limitless faith in the future and in progress found a ready home, at least intellectually, in the United States; but the American attitude toward modernism was not exactly one of open

Raymond Loewy Associates (American, established 1944). Studebaker Commander Starliner Coupe. 1953.
The Museum of Modern Art Archives, New York

arms. Perhaps because the nation had so little history of its own, Americans clung quite tenaciously to historical styles until the postwar period; it is important to remember that the Colonial Revival in interior decoration was taking place alongside the emergence of modernism during the 1930s and 1940s. It was left for functional modernism to enter the American home through the kitchen and the bathroom.[14] There is a deep irony here, in that European designers had looked to the hygiene and organizational efficiency of American kitchens and baths for inspiration following World War I. The rationalism they found there spread to the rest of the domicile, reflected notably in the Weimar Bauhaus's 1923 Haus am Horn project and Grete Schütte-Lihotzky's Frankfurt Kitchen of 1926. In the United States, however, the living room (formerly the "parlor") and the bedroom, to a lesser extent, proved highly resistant to the encroachment of International Style aesthetics until well after World War II. This is not the first of the many contradictions that characterize American design.

At midcentury, American design could be roughly divided into two camps: "functional" modernists, such as museum curators and design critics, versus what could be termed "stylistic" modernists, chiefly consisting of industrial and automotive designers. The functionalist group inherited its sense of aesthetics from academic European designs executed from the turn of the twentieth century to the early 1930s; the stylists were largely inspired by American culture (with an initial dose of French Art Moderne) and aimed to modernize the country while making a profit. Rather than following the Bauhaus dictum of "starting from zero," these stylists were quick to harness the public enthusiasm for streamlining and a vision of perpetual progress driven by mass consumption. Functionalists believed that people could be and *should* be taught to appreciate designs that abjured decoration and references outside of their "legitimate" roles as tools, whereas stylists felt the public had to be persuaded into an appreciation of modern forms, and believed, to cite just one example, that referring to aircraft details when designing an automobile was not illegitimate borrowing but, rather, cross-pollination.

When MoMA organized a symposium on automotive design in 1950, the notoriously inarticulate Harley Earl, vice president of General Motors' Styling Section, wanted no part of it (one wishes that he had sent his brash understudy, Bill Mitchell, as his emissary). It was Raymond Loewy—whose office designed the Communications Receiver (1947) for Hallicrafters Company and the Studebaker Starliner Coupe—who ended up defending Detroit against charges of aesthetic excess and was attacked from all sides. On the other side of the "stylistic" versus "functional" divide, George Nelson, who shaped Herman Miller's postwar design program—and designed the Mobile Table and the Tray Table (1948)—mockingly reviewed slides of stylistic details of Detroit products (to predictable howls of laughter) with no suggestion of how things might be improved (and continued, grudgingly, he assured others, to drive a Buick because of its dependability and creature comforts). Less than a year later, *Eight Automobiles* opened at the Museum, devoid of the products of General Motors, at a time when GM was so close to fifty percent of the market share in the United States that antitrust rumblings were heard in Washington. How could the American people be so wrong, their taste so misguided?

E. T. (Bob) Gregorie, Jr.
(American, 1908–2004).
Lincoln Zephyr. 1936.
The Museum of Modern
Art Archives, New York

E. T. (Bob) Gregorie, Jr.
(American, 1908–2004).
Edsel Ford
(American, 1893–1943).
Lincoln Continental. 1940.
The Museum of Modern
Art Archives, New York

Gordon Miller Buehrig
(American, 1904–1990).
Cord 810 Sedan. 1937.
The Museum of Modern
Art Archives, New York

Army Jeep. 1951.
The Museum of Modern
Art Archives, New York

Indeed, the American automobile is the elephant in the room when we talk about design in the United States. We taught the world how to mass produce the automobile, and motorized transport was a vital part of the American contribution to the Entente's victory in World War I. The automobile was the underpinning of the economic expansion of the 1920s, and Earl's 1927 La Salle convertible was the first mass-produced product to be successfully styled. The Jeep has been lauded many times as one of the most critical elements in the defeat of the Axis powers in World War II. Its straightforward stamped-steel aesthetic made it an acceptable candidate for inclusion in *Eight Automobiles*—both as a statement of undeniable functionalism and of what might be termed "anti-styling."[15] There are few more concise embodiments of postwar America in the popular imagination than the Cadillacs of the 1950s. Muscle cars captured a youth audience at the end of the 1960s and again at the end of the century. How did it become so easy to distinguish automotive design as somehow "different" from every other kind of design?

Part of that has to do with the designers themselves. When the Society of Industrial Designers (forerunner of today's Industrial Designers Society of America) was first organized in 1944, one of the principal stipulations for membership was that the candidate must demonstrate that he (and they were all men in the beginning) had produced designs that had gone into production in three different fields. Thus, stylists working on any number of vehicles during their careers in Detroit were excluded just as surely as in-house designers working solely on glassware for Corning. It privileged those who had founded the profession, and it may have, in the long run, deprived both industrial design and Detroit of a meeting ground for discussions that might have benefited both. Instead, there were shouting matches in the press and occasionally in public.

With Allied victory foreseeable at the end of 1943, the redesigning of the United States began in earnest. Fearing another economic recession, the federal government poured money into plans for prefabricated, mass-produced housing to be manufactured in the giant complexes, such as Ford's Willow Run Plant in Michigan, that had been built during the war to assemble aircraft. For the most part, these plans came to naught, and pent-up demand after nearly a decade and a half of austerity buoyed the economy and made government intervention on such a scale redundant. "By 1947," one designer of that generation told me, "we'd redesigned everything that could be redesigned."[16] The Flint line of kitchen tools manufactured by Ekco Products (1943–46), including the Flint Spatula, took such items out of the age of rust and made them requisite in most American homes. John Carroll designed the ubiquitous Presto Cheese Slicer around 1944, while aluminum was still strictly rationed. William H. Miller, Jr.'s somewhat provisional-looking Chair (c. 1944) for Gallowhur Chemical Corporation incorporated an inflatable polyvinyl chloride bag no doubt inspired by those used in lifesaving vests and inaugurated an era in which seating designs seemed to celebrate a do-it-yourself aesthetic no matter how carefully the details of their fabrication may have been sweated; see Side Chairs by Ray Komai (1949) and Donald R. Knorr (1948–50), for example. The superior glues and plastics that were developed as part of the Army Air Corps's Glider Program became the backbone of a revolution in plywood technology that fascinated Charles and Ray Eames, George Nelson, Eliot Noyes, and many others.

But even if this period has been dubbed the Eames Era by thousands of eBay users, numerous postwar designs were not the work of master designers but of engineers, inventor innovators, and those figures who might more typically be described as interior designers. William J. Russell designed a simple and understated, yet distinctive, flatiron—the Universal Electric Iron (before 1948), a company design for Landers, Frary & Clark—during a career in the electrical manufacturing industry bracketed by patents for heating coils and snap-action switches. Jay Monroe had the inspired idea of coupling the twelve-volt bulbs used in automobiles with a small transformer to make a lamp, and dubbed it the Tensor. There are notable examples in MoMA's collection of similar high-intensity lamps—the Koch Creations Lampette Reading Lamp and Michael Lax's Lytegem High-Intensity Lamp (1965), for example—and Monroe is represented here by another of his innovations: the Disposable Flashlight (1967). Freda Diamond certainly ranks among the top designers in MoMA's collection if sheer volume of sales is a factor. It is estimated that her Classic Crystal Glasses (1949) for Libbey Glass sold at least fifteen million units and they served as the standard barware for the Waldorf-Astoria Hotel in New York, surely one of the great case studies in reaching both ends of the market. T. H. Robsjohn-Gibbings's rather severe Table Lamp (1950) is among his least effusive designs and finds a meeting point between classicism and modernism, which sixteen years earlier might have made it a candidate for inclusion in *Machine Art.* Still, it is the aesthetic of the Eames Storage Unit (1950) that people associate with the 1950s in

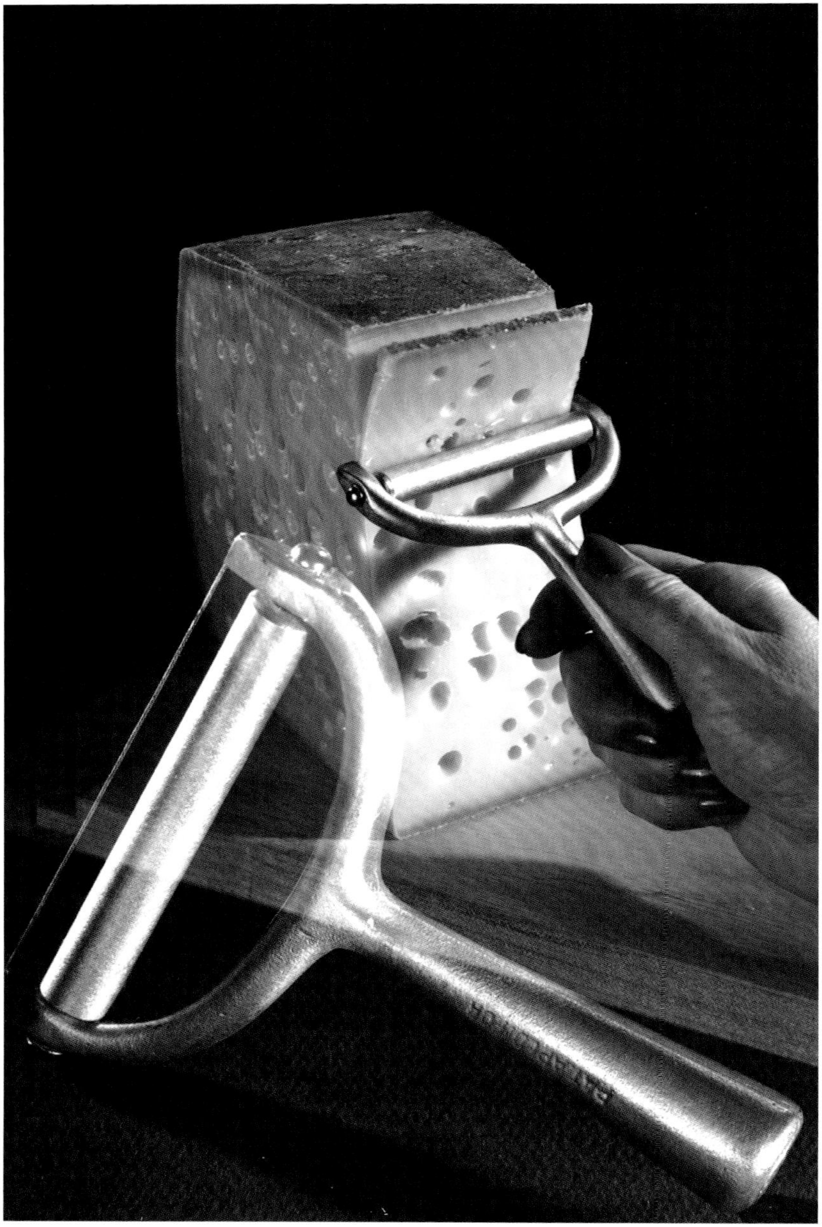

John Carroll
(American, 1892–1958).
Presto Cheese Slicer.
Date unknown.
Cast aluminum
and steel wire, 4½ x 3¾"
(11.4 x 9.5 cm).
Manufacturer:
R. A. Frederick Co.
The Museum of Modern
Art, New York. Gift of
Edgar Kaufmann, Jr., 1946

the United States. This is somewhat amusing, as both the unit's construction (complete with stamped angle brackets and guy wires) and aesthetic (modular construction) are very much pre–World War II. Not so Harry Bertoia's Armchair (1952) (far more elegant than similar designs from the Eames office), which seems to embrace a gridded, post-Einsteinian elastic space, nor Peter Hamburger's Hanging Light Structure (1966), which brings Buckminster Fuller and Kenneth Snelson's tensegrity structures to life, or at least into the home.[17]

It was a time when people put their silverware away and bought stainless-steel cutlery, and Russel Wright (who, one might say, started the whole idea of *not* polishing your silver) was there with his flowing, organic, and yet somehow quirky Highlight Flatware (1951). Don Wallance came up with the equally compelling Design 2 Flatware (1952). Just a few years later, he published his book *Shaping America's Products,* covering Wright's practice and Henry Dreyfuss's work for Bell Telephone Labs in a survey that mirrored designers' growing awareness of their significance in the shaping of corporate America. This is also when *Industrial Design* magazine was spun off from *Interiors.*[18] Next to such sophistication, there is a great deal of charm to the Frozen Food Knife made by W. R. Case & Sons Cutlery Company (1954). What material does frozen spinach most resemble? Probably something akin to soft pine, so the teeth of a crosscut saw blade found themselves grafted onto a body that looks a great deal like one in Ekco's line of Flint kitchen tools.

It is hard to imagine the 1950s, the decade of McCarthyism and the Red Scare, social upheaval over racial segregation, and the beginning of the Space Race, as an era at all characterized by complacency. But perhaps because of the sheer wealth available, the 1950s were a decade of both/and, exemplified by the cool perfection of such Mies-inspired pieces as Florence Knoll's Coffee Table (1954) and John Behringer's Link Bench (1961) and the anxious, agitated line of Saul Bass's 1955 poster for the film *The Man with the Golden Arm,* as well as the high-tech appearance of Eliot Noyes and Associates' IBM Dictating-Machine Stand (1956), which emphasizes its engineered components just as similar items designed by Sir Norman Foster would, decades later. In the 1960s, American designs begin to stand more strongly as individual sculptural statements—Richard Schultz's Petal Coffee Table (1960) and Lax's Lytegem High-Intensity Lamp, for example, and Henry Dreyfuss's Bell Telephone Laboratories Trimline Telephone (1960–65), which revealed a concern for the tactile frankly missing from his earlier designs.[19]

Ekco Products Co. (USA, established 1888). Candleholders. c. 1953. Spring wire, 1 x 2½" (2.5 x 6.3 cm). Manufacturer: Ekco Products Co., Chicago. The Museum of Modern Art, New York. Purchase, 1954

Pace Products (USA). Emergency Umbrella. c. 1955. Polyethylene film, wooden handle, and metal frame, open: 21 x 27½ x 27½" (53.3 x 69.9 x 69.9 cm); closed: 21 x 4 x 4" (53.3 x 10.2 x 10.2 cm). Manufacturer: Pace Products, Chicago. The Museum of Modern Art, New York. Gift of Arthur Drexler, 1956

The cultural upheaval of the late 1960s cannot be explored in detail in a work of this length, but Seymour Chwast's antiwar poster *End Bad Breath* (1967) captures the tone of the times by being humorous and subversive at the same time; it takes a moment to realize that Uncle Sam's halitosis stems from B-52 raids over North Vietnam, and the color scheme is not what we expect. In fact, it's the antithesis of red, white, and blue, so to speak: Stare at it steadily for a while and then look at a blank white wall. (Art Chantry's 1988 poster for the play *Propaganda* engages in a similar strategy by adopting the appearance of a three-dimensional image in its purposely out-of-register forms—viewers gradually become aware that they are, in fact, seeing the image properly, and, yes, it is confusing and somewhat painful, mirroring the subject's mental state.)

Koch Creations. USA. Lampette Reading Lamp (model E6). 1963. Plastic and metal, expanded: 16 x 3¼" (40.6 x 8.3 cm); compact: 6 x 3¼" (15.2 x 8.3 cm). Manufacturer: Koch Creations, Lynbrook, N.Y. The Museum of Modern Art, New York. Purchase, 1963

Bud Esry (American, born 1920). Saul Nesbitt (American, born 1925). Dax Frame. 1969. Injection-molded polystyrene and laminated box board, 7 x 5 x 1⁵⁄₁₆" (17.8 x 12.7 x 3.4 cm). Manufacturer: The Subsidiary Co. (division of Artmongers and Manufactory, Inc.), New York. The Museum of Modern Art, New York. Gift of the manufacturer, 1969

Bronislaw Zapolski (American, born 1917). Fluorescent Flashlight (model 10015). 1980. ABS polymer casing, clear styrene plastic bulb enclosure, and braided nylon wrist strap, 6⁵⁄₁₆ x 2⅛ x ¾" (16 x 5.4 x 1.9 cm). Manufacturer: Zelco Industries, Inc. The Museum of Modern Art, New York. Gift of the manufacturer, 1981

The subsequent dislocations of the 1970s brought the design profession its first intimations that it might be part of the problem, through its complicity in escalating consumer demand, rather than the perpetual solution to society's problems. It is highly ironic that just as human factors and ergonomic considerations became part of the typical working process of design offices, and professionals who had spent years, if not decades, working to assert their decidedly postwar sensibilities began to emerge from the shadows of big-name first-generation designers, industrial design found itself becoming more of a middle-management rather than executive-office concern. This was the legacy of the remarkable talent for selling that the first generation possessed; it is also a measure of its success that professional design had become the norm rather than the exception.[20]

Some extremely solid design was done in the United States in the 1970s, as an examination of the Dictaphone (1970) by William Lansing Plumb and Associates, Dale R. Caldwell and Paul D. Miller's Compact Super 8 Silent Movie Camera (1975), and Morison S. Cousins and Michael Alan Cousins's stunningly simple and elegant Promax Hair Dryer will attest. If the mechanical adopted the look of the organic as technology advanced, then by the 1970s it seemed that design could move beyond referencing one or the other exclusively. In other words, design broke free of being derivative of the natural or emulating the "objectivity" of the machine and began to operate within its own frame of reference. Peter Connolly's novel Glass Cutter (1980), with a grip that predisposes the user to apply it correctly; Greg Marting's Giro Atmos Bicycle Helmet (2003), which borrows its aesthetic from the net of mace that surrounds a nutmeg and yet is composed of only man-made materials with a wide range of densities; and Brian Alexander's Flo Cell Storage prototype, which mirrors its own zeitgeist of fractals and chaos theory just as surely as Frank Lloyd Wright's desk for the S. C. Johnson and Sons Administration Building (1936–39) captured the ethos of the Streamlined Decade: These objects not only look novel, but they even have a certain stylishness without being styled. Yet classic high modernism continued to be mined effectively even after its heyday, as seen in Nicos Zographos's eponymous Side Chair (1966), a convincing melding of Marcel Breuer's cantilevered Cesca Chair of 1928 with Giuseppe Terragni's 1932–36 Armchair for the Casa del Fascio in Italy.

The 1970s also saw more and more designers begin to market their innovations directly, as Clair H. Gingher and Alan Spigelman did with their respective scissors. Gingher's aim was to reduce the weight of a pair of dressmaker's shears while maintaining (actually improving) the consistency of fit between its cutting edges. He chose glass-fiber-reinforced nylon for the frame of the scissors because it does not distort under constant use, and cutting a single ounce from the typical three made the scissors easy to use for long periods. Spigelman was looking for a product to expand a line that included inexpensive but robust hairbrushes, mirrors, travel boxes, and other items in what could be a college student's tool kit. He created dishwasher-proof, oil-and-chemical-resistant scissors that seemed to acknowledge they would be misused at some point.

At this point I begin to feel the discomfort of an academic crossing over from the realm of fact into that of opinion, according to the rule that absolutely nothing can be considered "classic" until it is at least twenty-five years old. In many ways, the fact that the bright pastels and flying geometrics of the day haven't yet been successfully recycled indicates that public nostalgia for the 1980s is probably still restricted to outings to dance clubs. However, there is Ward Bennett's Double Helix Flatware (1985) to consider. It is highly representative of the decade while capturing the essence of modernist elegance through simplicity—and however stylish this flatware is, Bennett was concerned enough with function and his social role that he designed feeding utensils for physically impaired children around the same time.

As the 1980s waned and postmodern architecture moved from its radical to its institutional phase, industrial design seemed to regain its public esteem by harnessing human factors in ways that made its benefits a demonstrable "value-added" for consumers. When tools are designed with the physically impaired in mind, healthy people reap the benefits as well: Spend an hour peeling potatoes with the Ekco Vegetable Peeler (c. 1944) and then try the Good Grips Peeler (1989), the industrial design success story of its day (still in production), as a practical illustration of this principle. Probably no chair in recent memory has had the impact on the public imagination that Donald T. Chadwick and William Stumpf's Aeron Office Chair for Herman Miller (1992) has had; I believe it was the first ergonomic status symbol. However, the key to the Aeron story was Stumpf's realization, in watching a diminutive woman try to adjust a prototype of the chair to suit her small frame, that even with every trick in the book, one size does not fit all. The production chair was issued in three sizes in recognition of this fact, saying a great deal about the degree of mutual respect this outside design team and the Herman Miller staff had for one another. Rapidly shifting corporate landscapes don't usually inculcate such respectful relationships. A great design for a great client can evaporate with a single corporate acquisition or following the departure of an advocate on the business side of the relationship. And then, so to speak, design has to start all over again, and we return to the perpetual question: If it's a good design but it fails to have any impact in the marketplace, is it still a good design? Or is it simply fodder for future books about the history of design?

Morison S. Cousins
(American, 1934–2001).
Michael Alan Cousins
(American, born 1938).
Promax Compact Hairdryer.
1976. Lexan casing,
$6\frac{5}{8} \times 2\frac{5}{16} \times 6\frac{1}{4}$"
(16.8 x 5.8 x 15.9 cm).
The Gillette Company,
Boston, Mass.
The Museum of Modern
Art, New York.
Gift of the manufacturer,
1977

Brian Alexander (American, born 1963). Flo Cell Storage (prototype). 1997. Fiberglass, aluminum, steel, leather, and ABS, 24 x 37 x 20" (61 x 94 x 50.8 cm). Manufacturer: Haworth, Inc., USA. The Museum of Modern Art, New York. Gift of the manufacturer, 2001

This survey ends with two items that are typical of the state of design in the United States in the first decade of the twenty-first century: Deborah Adler and Klaus Rosburg's Target CleaRx Prescription System (2004) and Benjamin Rivera's Wave Multi-Tool for Leatherman (2004). Adler began to develop the graphics-driven design as a student project, and it has been celebrated in a somewhat "Yes, we can!" fashion within the design community, partly because it has a great backstory, just like the Good Grips Peeler. The Multi-tool (everyone who uses one, no matter which model, refers to it exclusively as "the Leatherman") is a solid improvement on an earlier version, which reimagined the Swiss Army Knife built around a pair of pliers rather than a cutting edge. It is extremely popular among American troops serving in Iraq. That says something not just about the challenges they face but about the equipment they typically have to operate with. It's one hell of a way to build a consumer base!

Anyone brave enough to foretell the future should read as a cautionary tale the predictions made in the popular literature of the 1940s and 1950s about the benefits that computers (invariably referred to as "electronic brains") would bring. In the years to come, the word "implants" may connote digital aids rather than silicone, and no doubt there will be benefits to accrue in designing equipment to deal with the disabilities encountered by an increasingly obese America. But many things will not change. People may get bigger, for example, but their hands will remain relatively stable in size; not all tools are rendered obsolete by the march of technology, and surely the heft and balance of a hammer will matter more than the material of its shaft in the hand of a skilled carpenter. A good machinist is known for the condition and efficacy of his tools, not their number; may we all try to be good machinists in this sense as we prepare for the Next Big Thing.

The great hope for the future of American design lies in our do-it-yourself approach, combined with the failure of the gatekeepers of technology to latch the barn door. The greatest danger lies in a situation in which design, having moved beyond styling, is threatened with subsumption under fashion. The sentiment probably did not originate with Henry Dreyfuss, but we might be wise to heed his comment to his fellow designers: "I'd rather be good than be original."

1. For example, in Paul Denison, "Christopher Dresser: A Design Revolution," *Journal of Design History* 18 (4) (2005): 387–90.

2. Tilmann Buddensieg makes this claim in the excellent book *Industriekultur: Peter Behrens and the AEG, 1907–1914,* by Buddensieg and Henning Rogge, trans. Ian Boyd Whyte (Cambridge, Mass.: MIT Press, 1984).

3. For an examination of these issues that stresses the role of the private sector in shaping mass production, see Donald R. Hoke, *Ingenious Yankees: The Rise of the American System of Manufactures in the Private Sector* (New York: Columbia University Press, 1990).

4. But engineers, at least at the top of the profession, were not immune to the demand for "good form." Engineers at Brown & Sharpe compiled an in-house manual of standard practices regarding aesthetics in the manufacture of machine tools. See Tim Putnam, "The Theory of Machine Design in the Second Industrial Age," *Journal of Design History* 1 (1) (1988): 24–34. Paul Jordan, a contemporary of Peter Behrens who directed marketing for the AEG, wrote, "A motor must look like a birthday present." Buddensieg and Rogge, *Industriekultur,* 110.

5. Raymond Loewy, *Never Leave Well Enough Alone* (New York: Simon and Schuster, 1951), 10.

6. Arthur Pulos, *American Design Ethic: A History of Industrial Design to 1940* (Cambridge, Mass.: MIT Press, 1983), 304.

7. Mies added his mother's maiden name, van der Rohe, for its upper-class connotations after World War I, and Jeanneret began referring to himself as Le Corbusier around 1920.

8. The illustration was originally published in *Life* magazine on April 19, 1917, at the time of the United States' entry into World War I. Gibson reused the image when he became head of the United States Committee on Public Information's Division of Pictorial Publicity.

9. Johnson was aided by the Museum's director, Alfred H. Barr, Jr., in organizing *Machine Art* and by Columbia University architectural historian Henry-Russell Hitchcock on the earlier show.

10. Henry Dreyfuss, *Designing for People* (New York: Simon and Schuster, 1955), 200.

11. Noyes was from a background that stressed hard work, integrity, and social duty (Teddy Roosevelt's "muscular Christianity") and could be highly moralizing about objects—part of the strain of Calvinism that is referred to as puritanism in the United States, with its suspicion of worldly goods, or at least of ostentation.

12. Actually, it shows up at a price of ten dollars in the exhibition's checklist.

13. One wonders if Noyes might have written the copy, as its tone is almost puritanical at times: "It is our understanding that the Japanese nightclubs have been closed for five years. That dancing is forbidden even in private Japanese homes. That one brand of shaving soap is sufficient for this need in Germany. That German women do not wear cosmetics. That large-scale spectator sports have been abandoned in Germany. If the Axis civilians can part with many peacetime luxuries, for the sake of their soldiers, we think American civilians can do that job better too." The Museum of Modern Art, *Useful Objects in Wartime Under $10* (New York: MoMA, 1942), 7.

14. Ellen Lupton and J. Abbott Miller's *The Kitchen, the Bathroom, and the Aesthetics of Waste: A Process of Elimination* (New York: Kiosk, 1992) does a remarkable job of linking the visual changes in the domestic environment with the literature of the times, most notably with Roy Sheldon and Egmont Arens's *Consumer Engineering: A New Technique for Prosperity* (New York: Harper & Bros., 1932).

15. Arthur Drexler found an apt phrase when he wrote that the Jeep has the "combined appeal of an intelligent dog and a useful gadget" in the catalogue for *Eight Automobiles*. Although the show was conceived prior to his arrival, it was the first exhibition he organized for the Museum. *Ten Automobiles* of 1953 demonstrated a cohesion in its selections missing from its predecessor, a tribute to Drexler's curatorial skills (or at least of his admiration for Pininfarina). He held the position of curator of design until 1956, when he became director of the Department of Architecture and Design, a title he retained until shortly before his death, in 1987. The disposable Emergency Umbrella (page 36) was one of his gifts to the collection, and he famously installed Buckminster Fuller's octet truss, tensegrity mast, and geodesic radome in the Museum's sculpture garden in 1959. His 1975 exhibition of architectural drawings from the École des Beaux-Arts—*The Architecture of the École des Beaux-Arts*—proved to be one of the pivotal points in the development of postmodernism.

16. Raymond Spilman, interview with the author, 1991.

17. Tensegrity is a structural system made famous by Fuller in which compression is discontinuous and tension is continuous. It has yet to find a practical structural application; any engineer will insist that additional strength can be built into a beam at less cost than using a tensegrity system for a comparable role.

18. The first issue's cover was designed by Alvin Lustig, and it featured the Studebaker Starliner Coupe, by Raymond Loewy Associates, as well as excerpts from Wallance's book.

19. This is largely because the project was under the guidance of a new generation of designers for whom the competition was seen not just as national but international; the positive feedback loop between Italy and the United States was never stronger nor more competitive than at this moment. On functional terms, the Trimline beats the pants off the 1965 Grillo Folding Telephone by Marco Zanuso (Italian) and Richard Sapper (German).

20. In fact, Teague's office was the first to advise clients on establishing in-house design services, in a way spelling the end of the consultant designer/corporate officer relationship that had characterized industrial design in the 1930s, 1940s, and 1950s.

CORONA
CORONA TYPEWRITER COMPANY, Inc.
3

the_beginning

1800_1939

Charles Stillwell

1845–1919

Flat-Bottomed Brown Paper
Grocery Bag, 1883
Paper, 10 x 5 x 3″
(25.4 x 12.7 x 7.6 cm)
Manufacturer: Duro Bag
Manufacturing Co., USA
Gift of the manufacturer, 2005

Samuel Gragg

1772–1855

Side Chair, c. 1808
Ash, oak, maple, and beech wood,
33 x 18½ x 29¼″
(83.8 x 47 x 74.3 cm),
seat h. 16¾″ (42.5 cm)
Manufacturer: Shop of Samuel Gragg, Furniture Warehouse, Boston
Marshall Cogan Purchase Fund, 1984

William Bradley

1868–1962

The Chap Book, Thanksgiving No., 1895
Zincograph, 20¾ x 14″
(52.7 x 35.6 cm)
Acquired by exchange, 1960

Corona Typewriter Co.

Established 1906

Corona Typewriter (model 3), 1912
Lacquered metal housing, unfolded: 6¼ x 10⅝ x 9¾″ (15.9 x 27 x 24.8 cm); folded: 4 x 10⅝ x 9″ (10.2 x 27 x 22.9 cm)
Manufacturer: Corona Typewriter Co., Inc., Groton, N.Y.
Gift of Wendy Goldhirsch, 2001

Louis Comfort Tiffany

1848–1933

Vase, c. 1900
Favrile glass, 6½ x 5⅛″
(16.5 x 13 cm)
Manufacturer: Tiffany Glass
& Decorating Co., New York
Joseph H. Heil Fund, 1965

George Ohr

1857–1918

Pitcher, c. 1900
Ceramic, 4⅛ x 6⅛ x 3¼″
(10.4 x 15.5 x 8.2 cm)
Gift of Charles Cowles and
Mary and David Robinson, 1983

Gustav Stickley

1858–1942

Armchair, c. 1907
Oak and leather, 29 x 26 x 27⅝″
(73.6 x 66 x 70.2 cm)
Manufacturer: Craftsman
Workshops, Eastwood, N.Y.
Gift of John C. Waddell, 1993

Frank Lloyd Wright

1867–1959

Office Armchair, 1904–06
Painted steel and oak,
36½ x 21 x 25″
(92.7 x 53.3 x 63.5 cm),
seat h. 19″ (48.2 cm)
Manufacturer: The Van Dorn Iron
Works Co., Cleveland, Ohio
Gift of Edgar Kaufmann, Jr., 1948

Charles Dana Gibson

1867–1944

U.S. Navy, "Here he is, Sir." We need him and you too! Navy Recruiting Station, 1917
Lithograph, 33 x 27"
(83.8 x 68.6 cm)
Printer: Latham Litho. & Print. Co., New York
Gift of Abby Aldrich Rockefeller, 1940

Walter Von Nessen

1889–1943

Desk Lamp, 1927
Chrome-plated brass, 16 x 24½"
(40.6 x 62.2 cm)
Manufacturer: Nessen Studio, Inc. (now Nessen Lamps, Inc.), New York
Gift of the manufacturer, 1956

Aluminum Cooking Utensil Co.

Established 1901

Wear-Ever Rotary Food Press, 1932
Aluminum, steel, and wood, pestle: l. 10¾″ (27.3 cm), diam. 1¾″ (4.4 cm); sieve–frame: assembled h. 9″ (22.9 cm), diam. 11⅜″ (28.9 cm)
Manufacturer: The Aluminum Cooking Utensil Co., New York
Gift of Lewis & Conger, 1947

Chase Brass & Copper Co.
Established 1876
Russel Wright
1904–1976

Pitcher, c. 1927–41
Copper-plated metal, 6 1/16 x 8 1/8 x 6 3/16″
(15.4 x 20.6 x 15.7 cm)
Manufacturer: Chase Brass & Copper Co.,
Waterbury, Conn.
Gift of Edgar Kaufmann, Jr., 1941

John G. Rideout

1898–1951

Teakettle, 1936
Metal and wood, 8½ x 9 13/16"
(21.6 x 25 cm)
Manufacturer: Wagner Mfg. Co.,
Sidney, Ohio
Purchase fund, 1944

Corning Glass Works

Established 1851

Double Boiler, c. 1938
Borosilicate glass and steel, bottom pot:
5½ x 11 x 5¾" (14 x 27.9 x 14.6 cm);
top pot: 5⅞ x 11 x 5¾"
(15 x 27.9 x 14.6 cm); lid: 5⅞" (15 cm)
Manufacturer: Corning Glass Works,
Corning, N.Y.
Purchase, 1944

Sherman L. Kelly

1869–1952

Ice Cream Scoop, 1935
Cast aluminum, 7 x 1¾″
(17.8 x 4.4 cm)
Manufacturer: Roll Dippers, Inc.
(formerly Zeroll Co.), Maumee,
Ohio
Purchase Fund, 1956

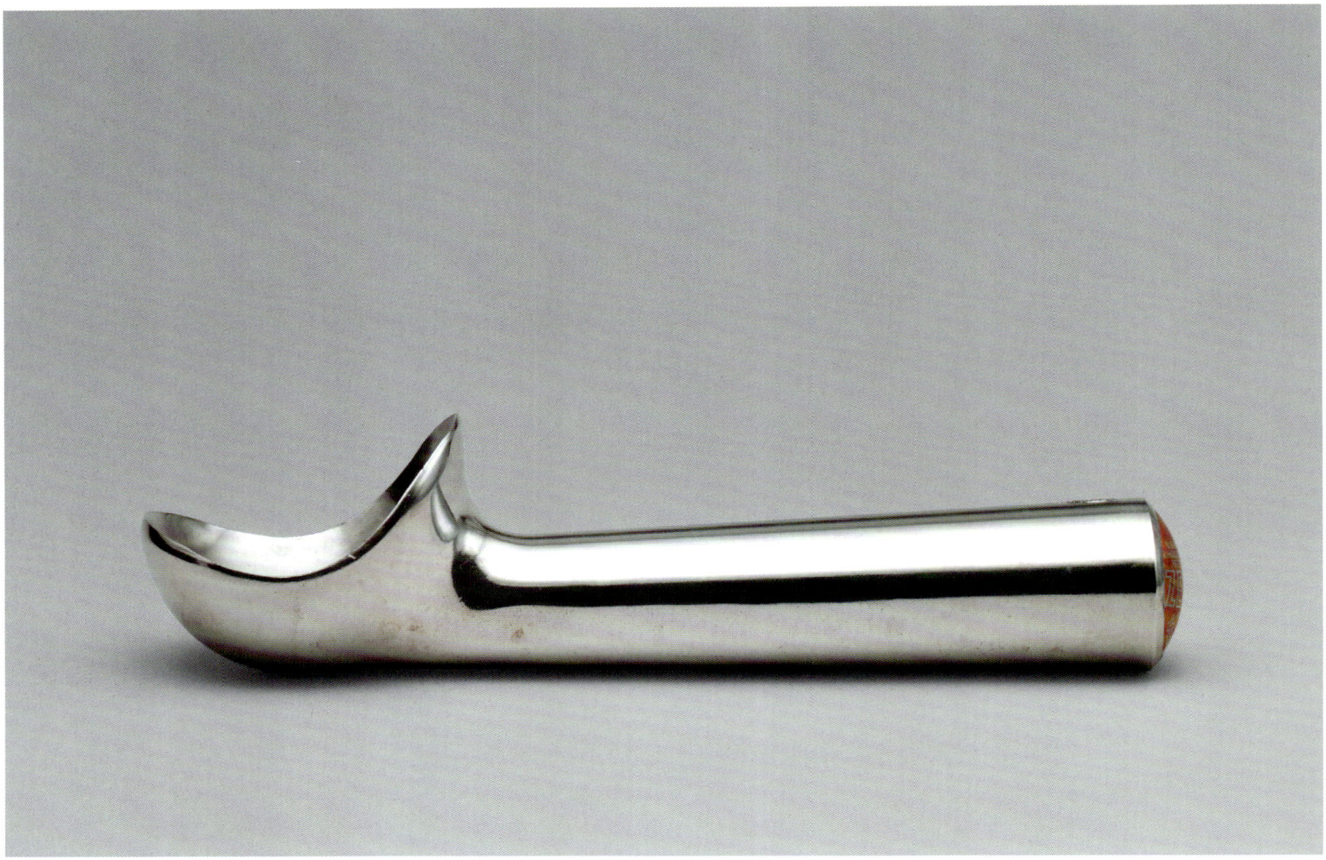

Russel Wright

1904–1976

"American Modern" Dinnerware, 1937
Ceramic, bowl: h. 2″ (5.1 cm), diam. 5 9/16″ (14.1 cm);
small plate: h. 9/16″ (1.4 cm), diam. 6 1/8″ (15.6 cm);
luncheon plate: h. 3/4″ (1.9 cm), diam. 8″ (20.3 cm);
dinner plate: h. 13/16″ (2.1 cm), diam. 9 13/16″ (25 cm);
water pitcher: h. 10 3/4″ (27.3 cm), diam. 6 1/2″ (16.5 cm);
teapot: h. 4 1/4 x w. 8 1/2″ (h. 10.8 x w. 21.6 cm)
Manufacturer: Steubenville Pottery Company,
Steubenville, Ohio
Gift of the manufacturer, 1944

McDonnell Aircraft Corp.

Established 1939

Gun Turret Seat, c. 1939–44
"Structomold" (plastic-impregnated laminated paper), 5¼ x 18 5/16 x 13" (13.3 x 46.5 x 33 cm)
Manufacturer: McDonnell Aircraft Corp., St. Louis, Mo.
Gift of the manufacturer, 1948

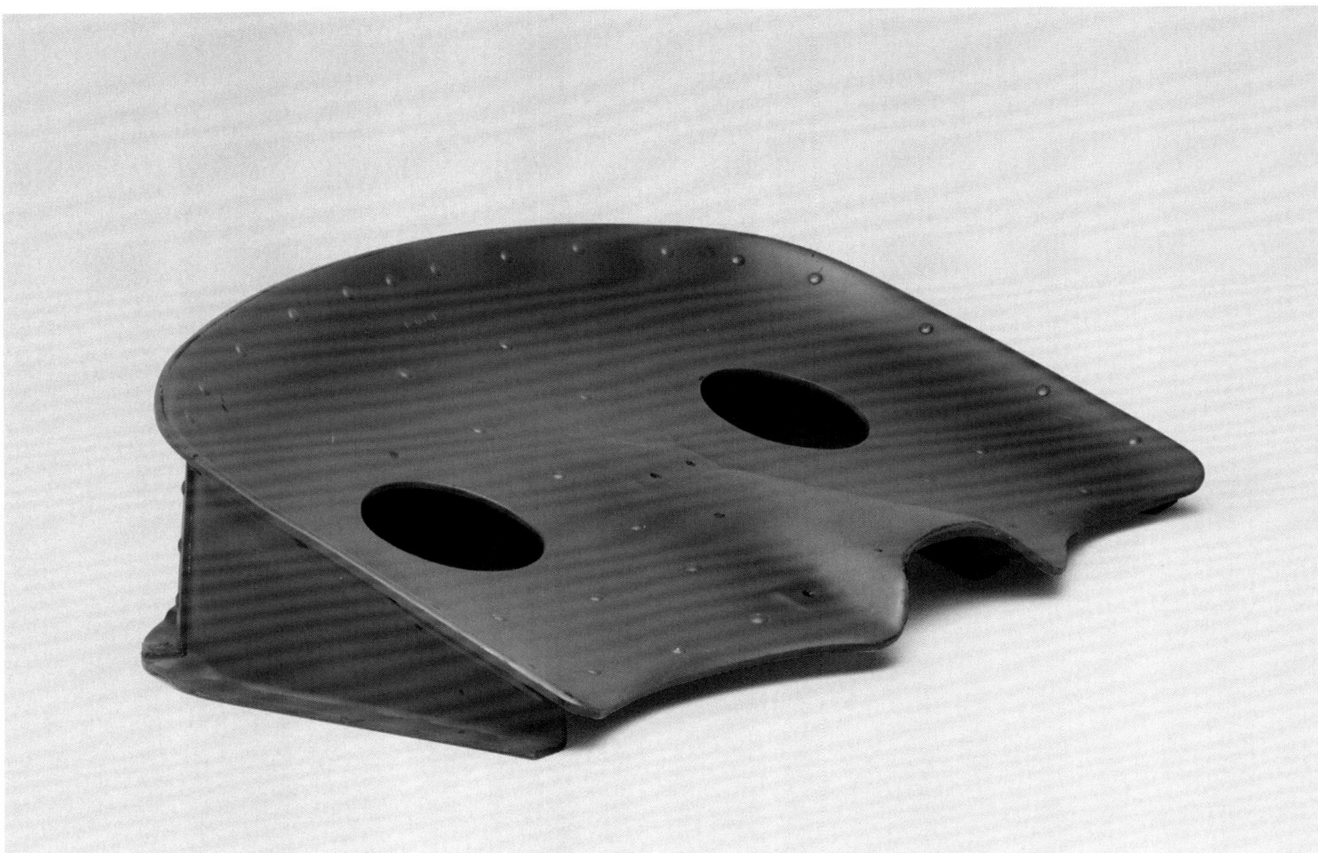

W. O. Langille

Born 1895. Death date unknown

Ribbonaire Fan, c. 1935
Bakelite and ribbon, 10 x 4½ x 7″
(25.4 x 11.4 x 17.8 cm)
Manufacturer: Diehl
Manufacturing Co., Elizabeth, N.J.
Gift of Singer Sewing Machine Co.,
1942

Isamu Noguchi

1904–1988

Radio Nurse Speaker, 1937
Bakelite resin, h. 8 ¼″ (21 cm), diam. 6 ½″ (16.5 cm)
Manufacturer: Zenith Radio Corp., Chicago
Gift of the designer, 1977

Lester Beall
1903–1969

Radio—Rural Electrification Administration, 1937
Silkscreen, 40¼ x 30″
(102 x 76.2 cm)
Gift of the designer, 1937

Lurelle Guild

1898–1986

Wear-Ever Teakettle (model 1403), c. 1932–33
Aluminum with plastic handle and lid knob,
8½ x 9½″ (21.6 x 24.1 cm)
Manufacturer: The Aluminum Cooking
Utensil Co., New York
Gift of the designer, 1944

Coors Porcelain Co.

Established 1910

Acid Pitcher, c. 1930
Glazed porcelain, 10¼ x 13½" (26 x 34.3 cm)
Manufacturer: Coors Porcelain Co., Golden, Colo.
Gift of the manufacturer, 1950

The Ironrite Ironer Co.

Established 1911

Health Chair, 1938
Steel and lacquered plywood,
26¼ x 17⅛ x 19½"
(66.7 x 43.5 x 49.5 cm)
Manufacturer: The Ironrite
Ironer Co., Detroit
Gift of the manufacturer, 1940

Frank Lloyd Wright

1867–1959

Desk, 1936–39
Wood and painted metal, 33¾″ x 7′ x 32″
(85.7 x 213.4 x 81.3 cm)
Manufacturer: Metal Office Furniture Co.
(now Steelcase, Inc.), New York
Purchase. Lily Auchincloss Fund, 1969

ZENITH

forties_and_fifties

1940_1959

Carl Anderson

1903–1989

Ross Bellah

1907–2004

Sectional Chair, c. 1940
Rattan with cotton-upholstered pads,
27 x 24½ x 26″ (68.6 x 62.2 x 66 cm)
Manufacturer: Ficks Reed Co.,
Cincinnati, Ohio
Purchase Fund, 1948

Richard Kelly

1910–1977

Table Lamp, c. 1940
Aluminum, steel, oak base, paper shade, and reflector bulb, 17¾ x 14″ (45.1 x 35.5 cm)
Manufacturer: Kelly & Thompson, USA
Gift of the designer, 1942

Russel Wright

1904–1976

Highlight Flatware, 1951
Stainless steel,
salad fork: l. 6¾″ (17 cm);
dinner fork: l. 7″ (17.8 cm);
dinner knife: l. 8¾″ (22.3 cm);
butter knife: l. 6⅜″ (16.2 cm);
tablespoon: l. 6⅞″ (17.4 cm);
teaspoon: l. 6¼″ (15.9 cm)
Manufacturer: John Hull Cutlers Corp., New York
Given anonymously, 2001

Jens Risom

Born Denmark, 1916

Low Lounge Chair (model 650), 1941
Maple and leather, 36 x 11 x 20″ (91.4 x 27.9 x 50.8 cm)
Manufacturer: Knoll Associates, New York
Gift of the designer, 2001

Charles Eames

1907–1978

Ray Eames

1912–1988

Lounge Chair, c. 1944
Molded plywood and steel rod,
28¾ x 30⅛ x 30″
(73 x 76.5 x 76.2 cm)
Gift of the designers, 1973

Collins Tools

Established 1826

Ax, c. 1946
Painted steel and wood handle,
overall: 27 13/16 x 6 3/8 x 1″
(70.6 x 16.2 x 2.5 cm);
head: 4 x 6 3/8 x 1″
(10.2 x 16.2 x 2.5 cm)
Manufacturer: Collins Tools
Gift of Abercrombie & Fitch, 1947

Raymond Loewy Associates

Established 1944

Communications Receiver (model S-40A), 1947
Steel casing, $8\frac{7}{8} \times 18\frac{1}{2} \times 9\frac{5}{8}$″
(22.5 x 47 x 24.5 cm)
Manufacturer: The Hallicrafters Co., Chicago
Gift of the manufacturer, 1948

Donald R. Knorr

Born 1922

Side Chair, 1948–50
Sheet metal, steel rods, rubber foam, and fabric, 30¼ x 23 x 19″ (76.8 x 58.4 x 48.3 cm)
Manufacturer: Knoll Associates, New York
Gift of the manufacturer, 1950

Ray Komai

Born 1918

Side Chair, 1949
Chrome-plated steel and molded wood, 29½ x 21 x 21″ (74.9 x 53.3 x 53.3 cm)
Manufacturer: J. G. Furniture Co., Brooklyn, N.Y.
Gift of the manufacturer, 1951

George Nelson

1908–1986

Tray Table (model 4950), 1948
Molded plywood and steel,
$19\frac{1}{2} \times 15\frac{13}{16} \times 15\frac{13}{16}$"
(49.5 x 40.2 x 40.2 cm)
Manufacturer: Herman Miller, Inc.,
Zeeland, Mich.
Gift of Fifty/50, 1984

Egmont Arens

1888–1966

Theodore C. Brookhart

1898–1942

Streamliner Meat Slicer (model 410), c. 1940
Aluminum, steel, and rubber, 13 x 20¼ x 17″
(33 x 51.4 x 43.2 cm)
Manufacturer: Hobart, USA
Gift of Eric Brill in memory of Abbie Hoffman, 1989

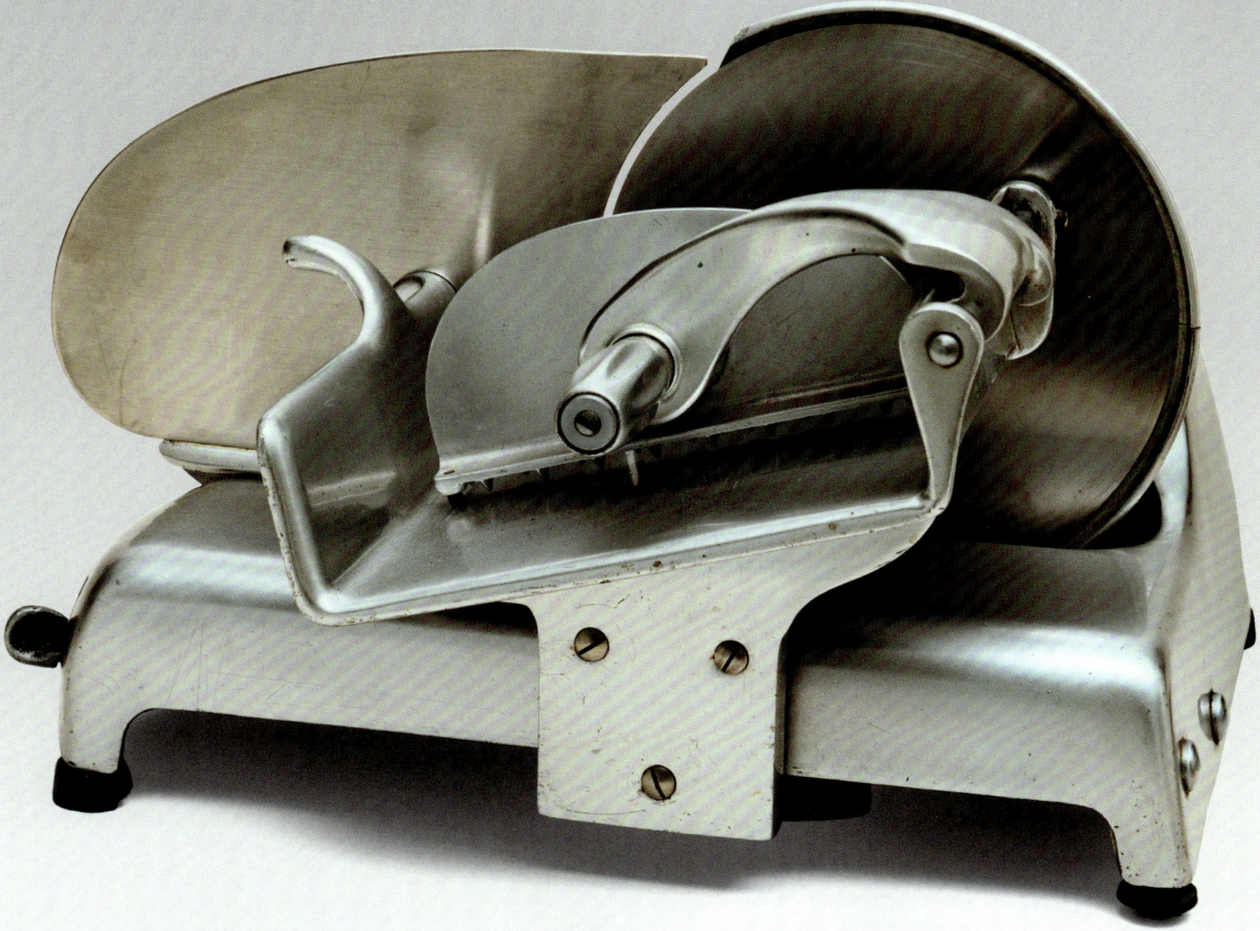

Ekco Products Co.

Established 1888

Flint Spatula, 1943–46
Stainless steel, phenolic plastic handle, and metal rivets, 13¾ x 3³⁄₁₆"
(34.9 x 8.1 cm)
Manufacturer: Ekco Products Co., Chicago
Gift of the manufacturer, 1947

Ekco Products Co.

Established 1888

Vegetable Peeler, c. 1944
Steel, 6⅞ x ⅞ x ½"
(17.5 x 2.2 x 1.3 cm)
Manufacturer: Ekco Products Co., Chicago
Purchase, 1956

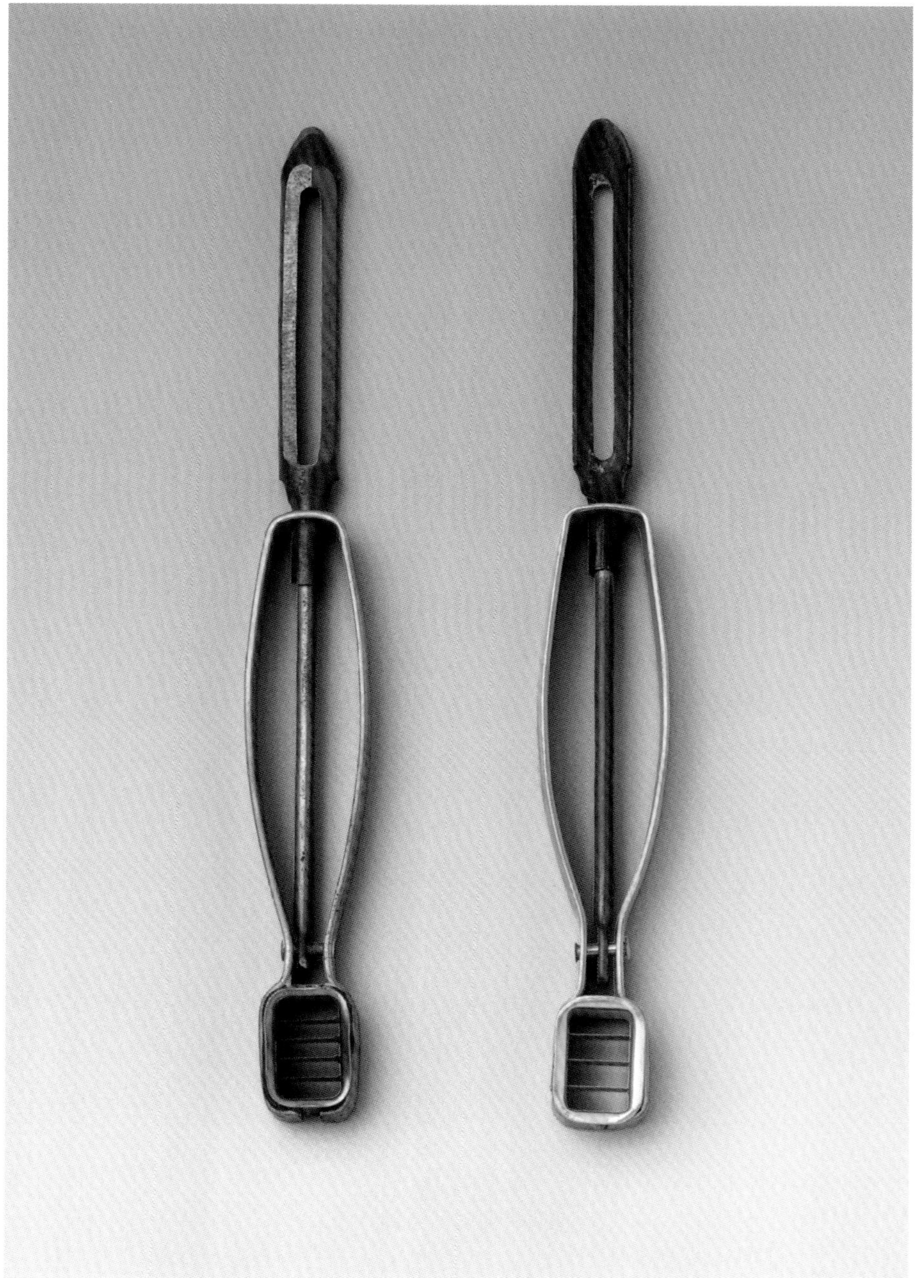

Landers, Frary & Clark

Before 1865

William J. Russell

Active 1934–1965

Universal Electric Iron, before 1948
Chrome and plastic, 4 x 9⅛″ (10.2 x 23.2 cm)
Manufacturer: Landers, Frary & Clark,
New Britain, Conn.
Gift of the manufacturer, 1948

Freda Diamond

1905–1998

Classic Crystal Glasses, 1949
Glass, left to right:
5 x 3 3/16" (12.7 x 8.1 cm);
4 5/8 x 3" (11.7 x 7.6 cm);
3 3/8 x 3 1/4" (8.6 x 8.3 cm);
3 1/2 x 2 5/8" (8.9 x 6.7 cm)
Manufacturer: Libbey Glass Company Division, Owens-Illinois Co., Toledo, Ohio
Given anonymously, 2001

Jack Heaney

Dates unknown

"Aluma-Stack" Stacking Chair, c. 1947
Painted aluminum frame and canvas seat, 30¼ x 16½ x 16⅞"
(76.8 x 41.9 x 42.9 cm)
Manufacturer: Treitel-Gratz Co., Inc., New York
Gift of Treitel-Gratz Co., Inc., 1948

Glenn Grohe

1912–1956

He's Watching You, 1942
Offset lithograph, 14⅛ x 10″
(35.8 x 25.4 cm)
Printer: US Government Printing Office
Gift of the Office for Emergency Management, 1968

Saul Bass

1921–1996

The Man with the Golden Arm,
1955
Lithograph, 40½ x 27"
(102.4 x 68.5 cm)
Gift of Otto Preminger Productions,
United Artists, 1956

Robert Budlong

1902–1955

Pacemaker Table Radio, 1948
Plastic and metal housing,
6⅞ x 12¼ x 6¾″ (17.5 x 31.1 x 17.1 cm)
Manufacturer: Zenith Radio Corp., Chicago
Gift of the manufacturer, 1950

Joseph Burnett

Born 1923

Three-Piece Table Lamp, c. 1950
Metal, brass, paper, and foil, overall: 16½ x 20½″ (41.9 x 52.1 cm);
stand: 16½ x 20½″ (41.9 x 52.1 cm);
husk: 8½ x 5½″ (21.6 x 14 cm);
reflector shade: 3 x 20″ (7.6 x 50.8 cm)
Manufacturer: Heifetz Mfg. Co., New York
Gift of the manufacturer, 1951

Charles Eames
1907–1978
Ray Eames
1912–1988

Eames Storage Unit (ESU), 1950
Plastic-coated plywood, lacquered Masonite, and chrome-plated steel, 58½ x 47 x 16¾″
(148.6 x 119.4 x 42.5 cm)
Manufacturer: Herman Miller, Inc., Zeeland, Mich.
Gift of John C. Waddell, 1992

Leo Collins

Dates unknown

Joseph Schlinger

Dates unknown

Paper Welder, 1954
Chrome-plated metal, 4½ x 1⅞ x 5¼″
(11.4 x 4.7 x 13.4 cm)
Manufacturer: Paper Welder, Inc., Medina, N.Y.
Gift of Bonniers, Inc., 1968

Harold Cohen

Dates unknown

Lounge Chair, 1951
Steel tube and woven fiber,
29 x 20 x 29¾″
(73.7 x 50.8 x 75.5 cm)
Manufacturer: Designers in Production, Chicago
Gift of the manufacturer, 1955

Harry Bertoia

Born Italy, 1915–1978

Armchair, 1952
Plastic-coated wire-and-rod frame and tweed upholstery over foam-rubber padding, 30½ x 35$\frac{1}{16}$ x 29″
(77.5 x 89.1 x 73.7 cm)
Manufacturer: Knoll Associates, New York
Gift of Knoll Associates, 1953

T. H. Robsjohn-Gibbings

Born England, 1905–1976

Table Lamp (model 170), 1950
Metal and linen, 21½ x 17″
(54.7 x 43.2 cm)
Manufacturer: Widdicomb
Furniture Co., New York
Gift of the manufacturer, 1952

Anthony Ingolia

Born 1921

Table Lamp, c. 1950
Steel, nickel, aluminum, and
enamel, adjustable height ranges
from min. h. 15″ (38.1 cm)
to max. h. 22″ (55.9 cm)
Manufacturer: Heifetz Mfg. Co.,
New York
Gift of the manufacturer, 1951

Philip Johnson

1906–2005

Richard Kelly

1910–1977

Floor Lamp, 1950
Brass and painted metal,
42 x 25″ (106.7 x 63.5 cm)
Manufacturer: Edison Price,
New York

William H. Miller, Jr.

Dates unknown

Chair, c. 1944
Vinylite (polyvinyl chloride) tube ring, plywood frame, aluminum legs, and string netting, 28 x 29½ x 31½″ (71.1 x 74.9 x 80 cm)
Manufacturer: Gallowhur Chemical Corp., Windsor, Vt.
Gift of the manufacturer, 1944
Gift of Philip Johnson, 1958

Eliot Noyes and Associates

Established c. 1956

IBM Dictating-Machine Stand,
1956
Walnut, masonite, and steel,
24⅛ x 17½ x 17½"
(61.3 x 44.5 x 44.5 cm)
Andrew Cogan Purchase Fund,
2002

W. R. Case & Sons Cutlery Co.

Established 1889

Frozen Food Knife, 1954
Hard rubber and stainless steel,
14½″ (36.8 cm)
Purchase, 1956

Don Wallance

1909–1990

Design 2 Flatware, 1956
Stainless steel,
dinner fork: 7″ (17.8 cm);
dinner knife: 7⅝″ (19.4 cm);
soup spoon: 6⅜″ (16.2 cm)
Manufacturer: H. E. Lauffer Co., Inc., West Germany (now Germany)
Gift of H. E. Lauffer Co., Inc., 1958

Florence Knoll

Born 1917

Coffee Table, 1954
Rosewood and chrome-plated metal, 17 x 27⅛ x 27⅛"
(43.2 x 68.9 x 68.9 cm)
Manufacturer: Knoll International, Inc., New York
Barbara Jakobson Purchase Fund, 1997

Celebration!
200 Years USA/50 Years CCA

sixties_and_seventies

1960_1979

Mort N. Marton

Born 1918

Beryl Marton

Dates unknown

Ice Bucket, 1963
Cork with milk-glass liner,
7⅝ x 8½ x 8½"
(19.4 x 21.6 x 21.6 cm)
Manufacturer: Mort N. Marton Designs, New York
Gift of David Whitney, 1963

Seymour Chwast

Born 1931

End Bad Breath., 1967
Offset lithograph,
37 x 24" (94 x 61 cm)
Gift of Push Pin Studios, 1969

Henry Dreyfuss

1904–1972

Bell Telephone Laboratories
Trimline Telephone, 1960–65
Materials unknown,
3½ x 8½ x 2¾″ (8.9 x 21.6 x 7 cm)
Manufacturer: Western Electric,
New York
Gift of the Bell Telephone System,
1965

Eliot Noyes and Associates

Established c. 1956

Executary Transcribing Machine (model 272), 1966
Plastic casing, 3⅛ x 10⅜ x 9⅞″ (8 x 26.3 x 25.1 cm)
Manufacturer: International Business Machines Corp., Armonk, N.Y.
Gift of IBM, New York, 1969

Danny Ho Fong

Born 1915

Wave Chaise, 1966
Wrought-iron and rattan
with cotton-upholstered pad,
16¼″ x 7′ x 24¼″
(41.2 x 213.3 x 61.6 cm)
Manufacturer: Tropi-Cal Co.,
Los Angeles
Gift of the manufacturer, 1967

David Rowland

Born 1924

40/4 Stacking Chair, 1964
Chrome-plated steel, vinyl-coated steel sheet, and plastic glides,
30 x 20 x 21″
(76.2 x 50.8 x 53.3 cm)
Manufacturer: The General Fireproofing Co., Youngstown, Ohio
Gift of the manufacturer, 1965

Peter Hamburger

Born 1941

Hanging Light Structure, 1966
Lumiline lamps, acrylic, and coated wire, 21½ x 18½ x 18½″ (54.6 x 47 x 47 cm)
Manufacturer: Peter Hamburger Designs, New York
Gift of the designer, 1969

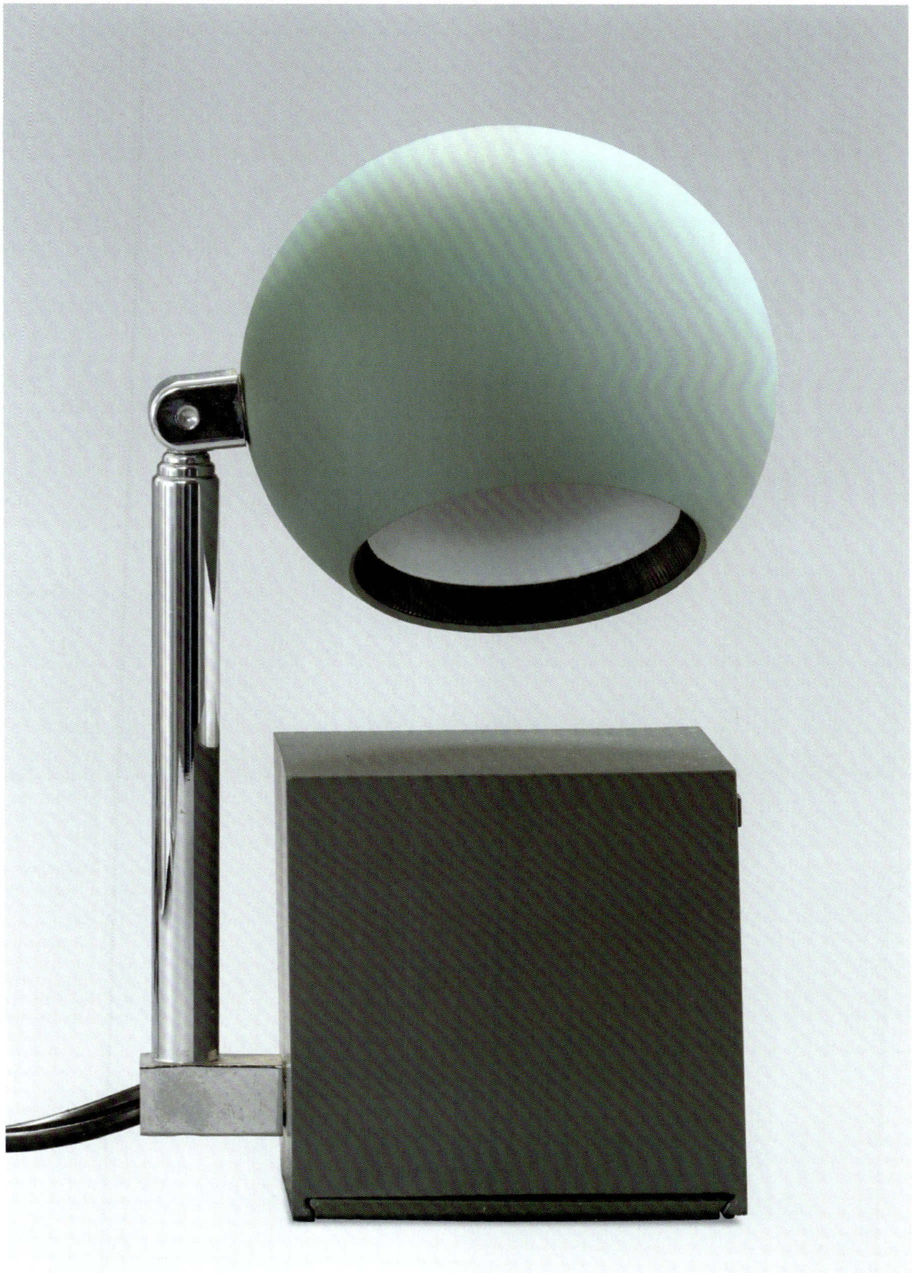

Michael Lax

1929–1999

Lytegem High-Intensity Lamp, 1965
Plastic, zinc, and aluminum, 15 x 3 x 3½″ (38.1 x 7.6 x 8.9 cm)
Manufacturer: Lightolier, Inc., Jersey City, N.J.
Gift of the manufacturer, 1965

Milton Glaser

Born 1929

The Sound is WOR-FM 98.7, 1966
Lithograph, 45⅛ x 58½"
(114.6 x 148.6 cm)
Gift of Station WOR-FM, 1966

Kenneth Brozen

1927–1989

Serving Bowl, 1963
Plastic, outer bowl: 8¼ x 9¾″ (21 x 24.8 cm), inner bowl: 7⅞ x 9¾″ (20 x 24.8 cm), lid: 5 x 7 11/16″ (12.7 x 19.5 cm)
Manufacturer: Robinson, Lewis and Rubin, Inc., Brooklyn, N.Y.
Gift of the designer, 2001

Jay Monroe

1926–2007

Disposable Flashlight, 1967
Plastic casing, 3½ x 2 x 1⅛″
(89 x 5.1 x 2.9 cm)
Manufacturer: Tensor Corporation, Brooklyn, N.Y.
Gift of the manufacturer, 1969

Harvey J. Finison

1916–1987

Bottle Opener, 1977
Stainless steel, l. 5″ (12.7 cm)
Manufacturer: Northampton
Cutlery Co., Northampton, Mass.
Gift of the manufacturer, 1978

Richard Schultz

Born 1926

Petal Coffee Table, 1960
Redwood, aluminum, and cast iron, 14⅞ x 42″ (37.8 x 106.7 cm)
Manufacturer: Richard Schultz Design, Palm, Pa.
Gift of the manufacturer, 2000

Nicos Zographos

Born 1931

Zographos Side Chair, 1966
Chrome-plated tubular steel
with leather, 31 x 20 x 22″
(78.7 x 50.8 x 55.9 cm)
Manufacturer: The General
Fireproofing Co., Youngstown, Ohio
Gift of the manufacturer, 1969

Clair H. Gingher

Born 1922

Scissors, 1979
Glass-filled nylon (chopped fiberglass and nylon) and stainless steel, 8⅛ x 2¾" (20.6 x 7 cm)
Manufacturer: Gingher, Inc., Greensboro, N.C.
Gift of the manufacturer, 1984

William Bonnell

Dates unknown

Celebration! 200 Years USA/ 50 Years CCA, 1976
Offset lithograph, 34 x 22″ (86.3 x 55.9 cm)
Printer: Mossberg Printing
Gift of the designer, 1981

Larry Bamford

Born 1937

Backpacker Hunting Knife, 1973
Stainless steel, 6⅛ x 1⅛ x ⅛"
(15.5 x 2.8 x 0.3 cm)
Manufacturer: Gro Knives, Inc., USA
Gift of the designer, 1974

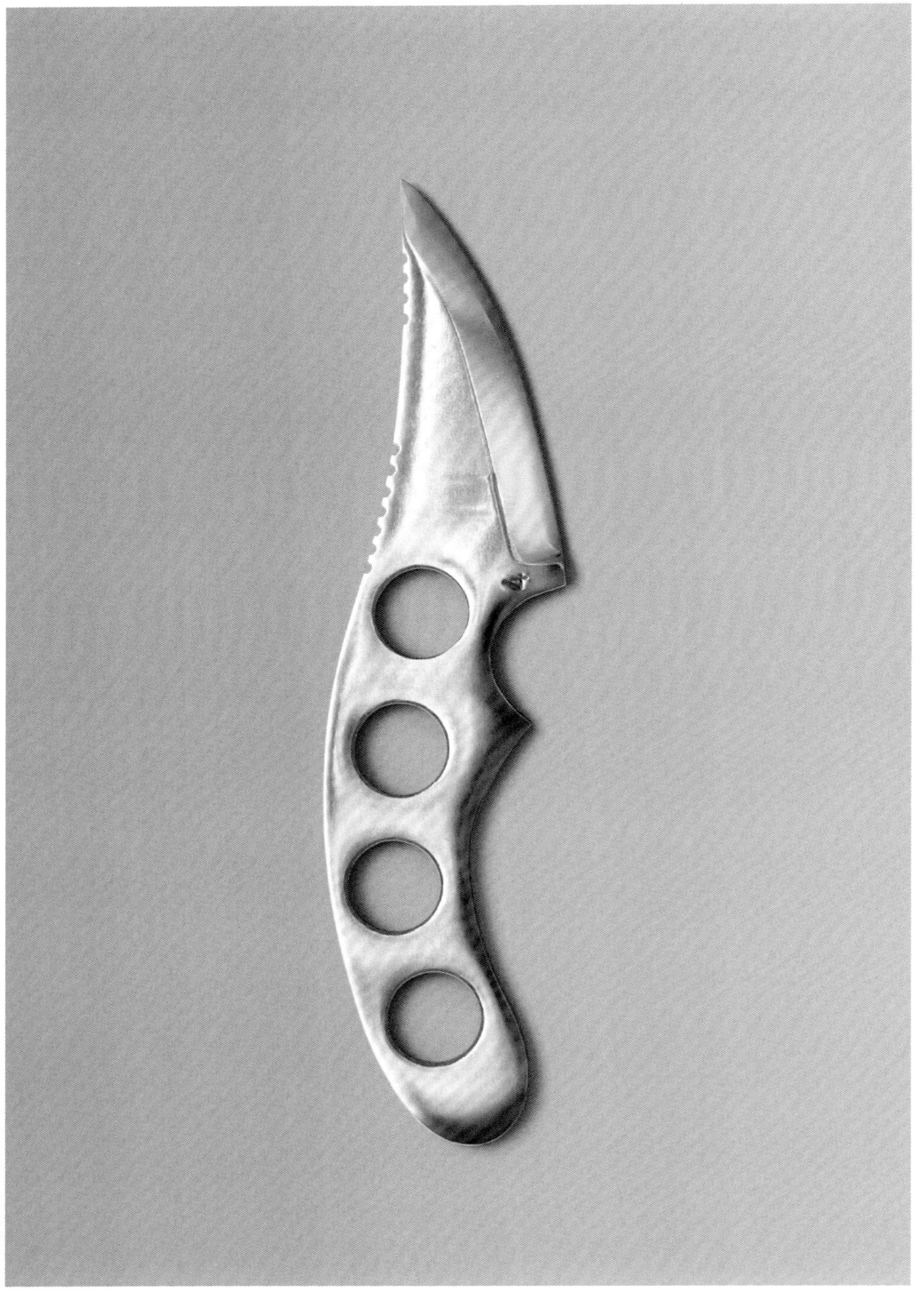

William Lansing Plumb
and Associates

Established 1963

Dictaphone (model 10), 1970
ABS polymer casing,
5 x 2½ x 1⅛″
(12.7 x 6.3 x 2.9 cm)
Manufacturer: Dictaphone AG,
Killwangen, Switzerland
Gift of the manufacturer, 1976

Frank O. Gehry

Born Canada, 1929

Easy Edges Body Contour Rocker, 1971
Laminated corrugated fiberboard,
26 x 24⅜ x 41¾"
(66 x 61.9 x 106 cm)
Gift of the manufacturer, 2002

Peter Danko

Born 1949

Armchair, 1976
One-piece molded plywood
and leather seat, 31½ x 21¾ x 24″
(80 x 55.2 x 61 cm),
seat h. 18½″ (47 cm)
Gottesman Fund, 1978

John Behringer

Born 1939

Link Bench (model 656), 1961
Chrome-plated steel and leather,
15¾ x 62 x 20″
(40 x 157.5 x 50.8 cm)
Manufacturer: Fabry Associates,
Inc., New York
Phyllis B. Lambert Fund, 1963

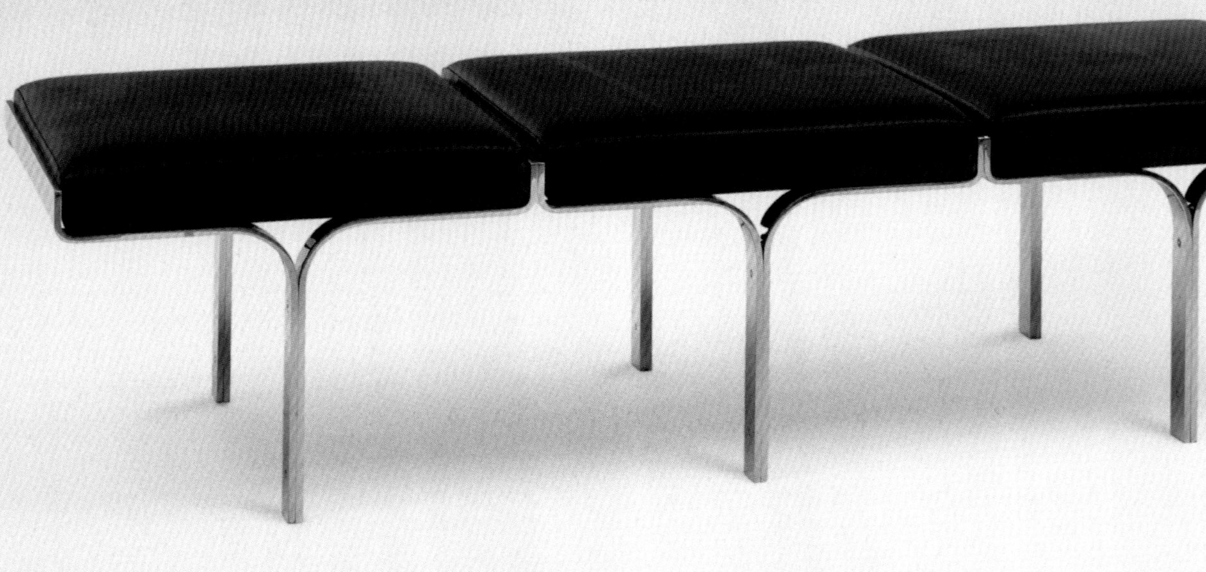

Dale R. Caldwell
Born 1929
Paul D. Miller
Born 1943

Compact Super 8 Silent Movie Camera (model 102A), 1975
ABS polymer casing, aluminum, rubber, and cloth strap,
3¼ x 1⅝ x 7¾" (8.3 x 4.1 x 19.7 cm)
Manufacturer: GAF Corporation, New York
Gift of the manufacturer, 1976

eighties_nineties_and_beyond

1980_2004

Henry Altchek

Born 1943

Bottle Opener, 1980
18/8 stainless steel with brushed satin finish, 7 x 1⅝ x ¹⁄₁₆"
(17.8 x 4.1 x 0.2 cm)
Manufacturer: Kaplan/Aronson, New York
Gift of the designer, 1980

Stephen Armellino

Born 1955

Bullet-Resistant Mask, 1983
Kevlar and polyester resin,
11 x 6¾ x 3¾"
(28 x 17.1 x 9.5 cm)
Manufacturer: US Armor Corp.
Gift of the manufacturer, 1992

Ward Bennett

1917–2003

Double Helix Flatware, 1985
Stainless steel,
salad fork: 6¾ x ⅞″ (17.2 x 2.3 cm);
fork: 7⅞ x 1″ (20 x 2.5 cm);
knife: 9⅛ x ½″ (23.2 x .3 cm);
soup spoon: 7½ x 1¼″ (19 x 3.2 cm);
spoon: 6½ x 1⅛″ (16.5 x 2.8 cm)
Manufacturer: Sasaki, Japan
Zaidee Dufallo Fund, 1986

Peter Connolly

Born 1939

Glass Cutter, 1980
Die-cast zinc with baked enamel finish and urethane coating,
1⅛ x 1¾ x 5¾″ (2.8 x 4.5 x 14.6 cm)
Manufacturer: Red Devil, Union, N.J.
Gift of the designer, 1981

Ben Winter

Born 1952

Zwirl Football, 1985
Polyurethane foam, h. 9″ (22.9 cm), diam. 5½″ (14 cm)
Manufacturer: Zwirl Sales Inc., Danville, Calif.
Purchase, 1992

Paul Rand

1915–1996

IBM, 1982
Offset lithograph,
36 x 24″ (91.5 x 61 cm)
Gift of the designer, 1983

Alan Spigelman

Born 1939

Scissors, 1980
Epoxy-coated stainless steel and plastic, 7 3/16 x 2 9/16"
(18.2 x 6.5 cm)
Manufacturer: Wings over the World Corp., New York
Gift of the designer, 1983

Art Chantry

Born 1954

Propaganda, 1988
Offset lithograph,
35⅛ x 21⅜" (61.5 x 46 cm)
Gift of the designer, 1995

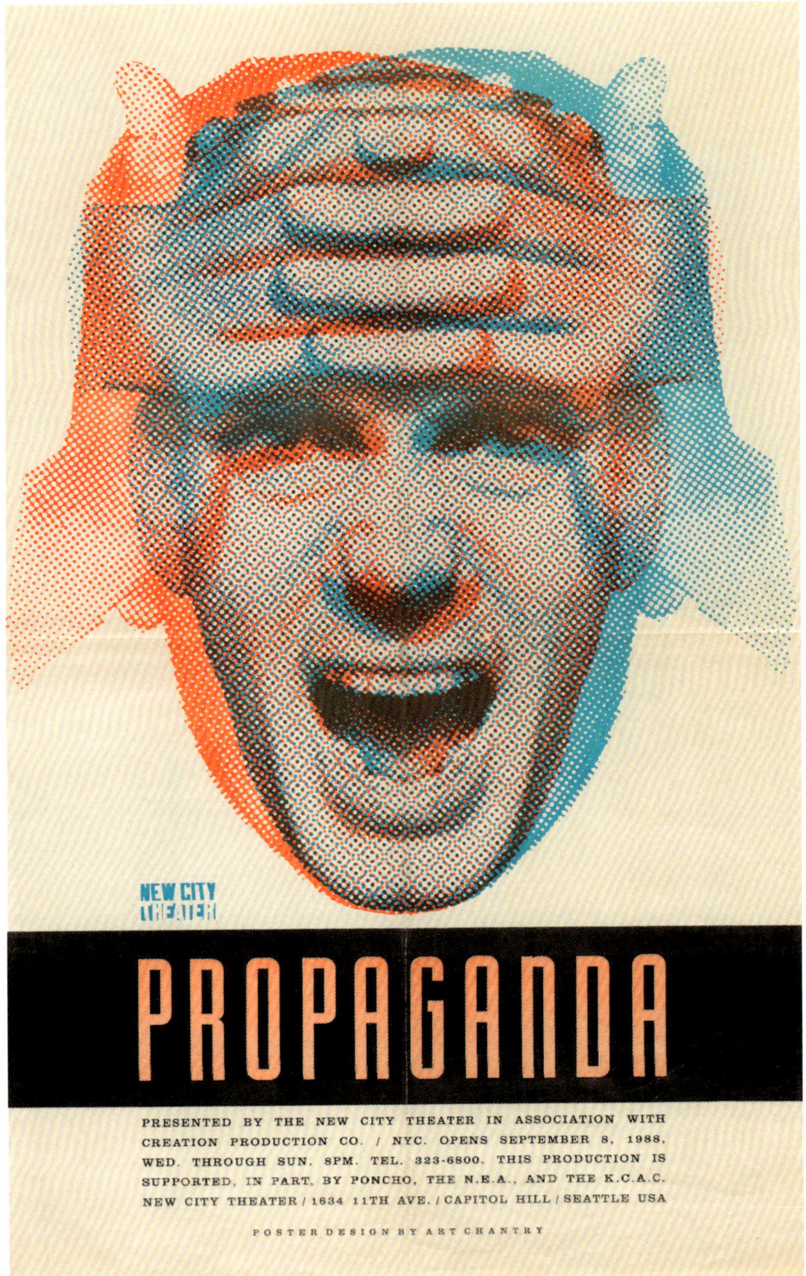

Smart Design

Established 1979

Good Grips Peeler, 1989
Stainless steel and rubber,
8¾ x 5 x 1″ (22.2 x 12.7 x 2.5 cm)
Manufacturer: Oxo International,
New York
Gift of the designers, 1994

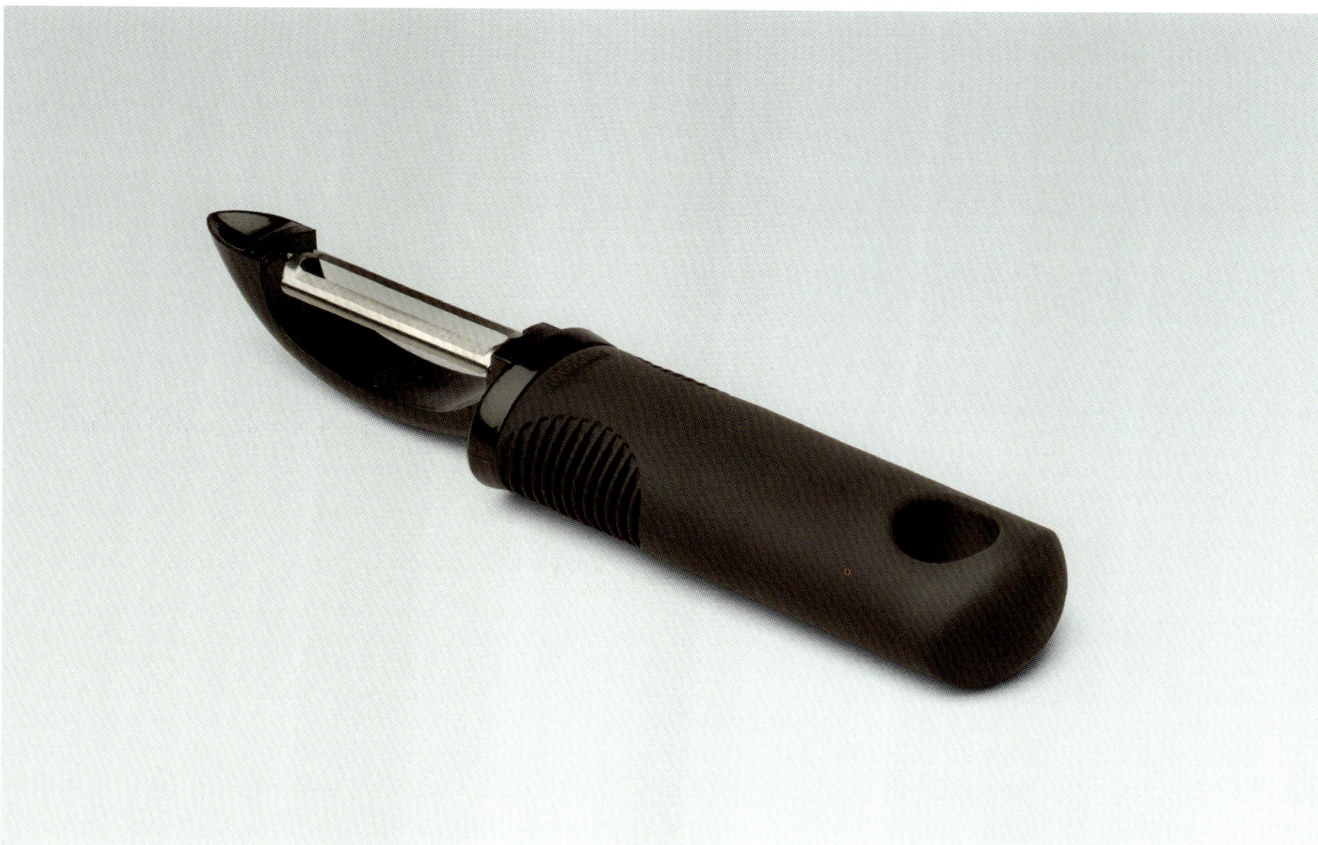

Nickie Campbell

Born 1962

William Campbell

Born 1954

Infant's Bottle, 1983
Polyethylene, large: 8 1/8 x 3 x 1 7/8"
(20.6 x 7.6 x 4.8 cm);
small: 5 5/8 x 2 9/16 x 1 3/4"
(14.3 x 6.5 x 4.5 cm)
Manufacturer: Ansa Bottle Company, Inc., Muskogee, Okla.
Gift of the manufacturer, 1988

Bruce Ancona

Born 1956

Louis Henry

Born 1971

Clip'n Stay Clothespins, 1998
Low-density polypropylene,
3¼ x 1½ x ⅝" (8.3 x 3.8 x 1.6 cm)
Manufacturer: Ekco Housewares,
Franklin Park, Ill.
Gift of the manufacturer, 2000

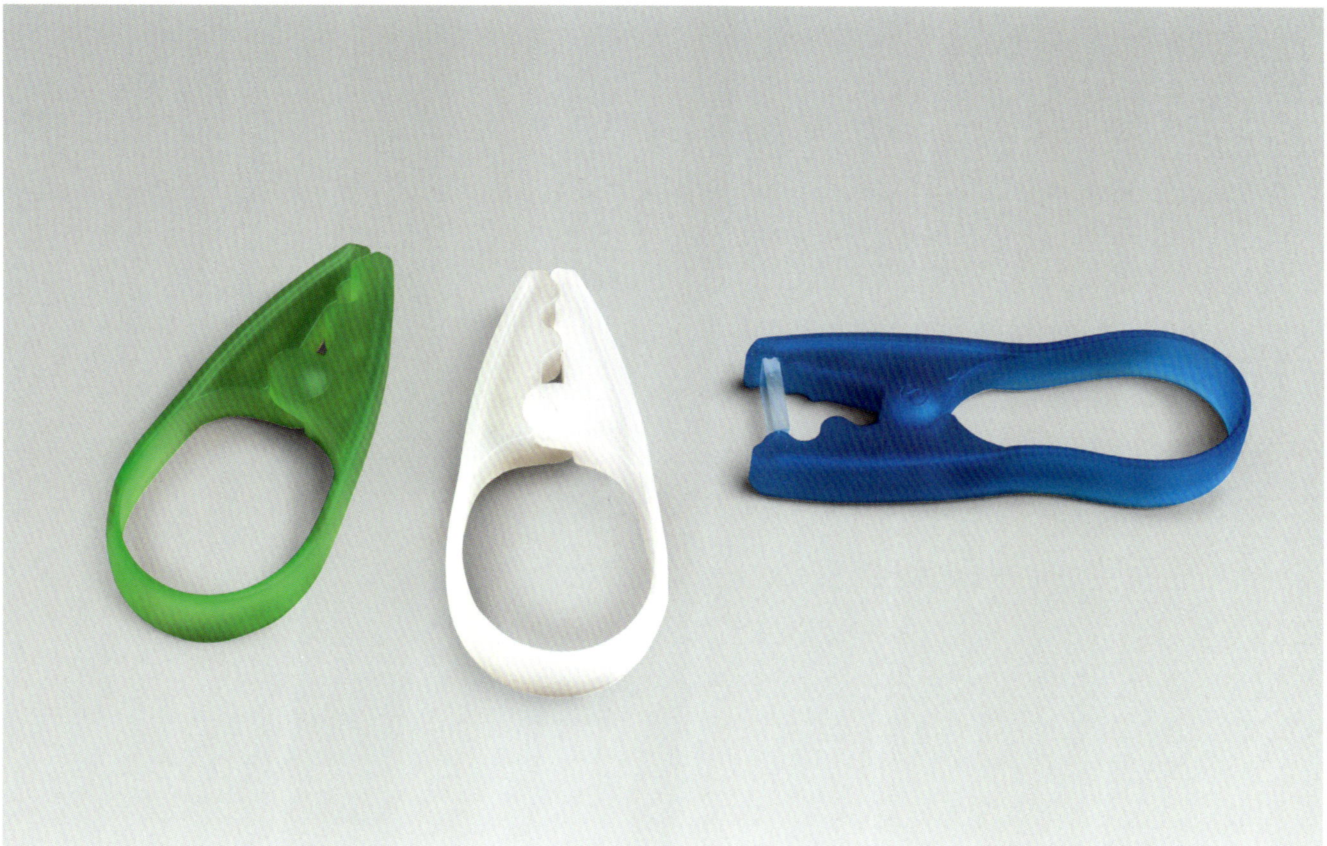

Donald T. Chadwick

Born 1936

William Stumpf

1936–2006

Aeron Office Chair, 1992
Structure: glass-reinforced polyester and die-cast aluminum; pellicle: Hytrel polymer, polyester, and Lycra; dimensions range from min. h. of 37¼″ (94.6 cm) to max. h. of 43″ (109.2 cm) x 28½ x 28½″ (72.4 x 72.4 cm)
Manufacturer: Herman Miller, Inc., Zeeland, Mich.
Gift of the employees of Herman Miller, 1994

IDEO

Established 1978

Paul Bradley

Born 1960

Lawrence Lam

Born 1960

3-D Mouse, 1991
Injection-molded ABS plastic,
2 7/8 x 5 3/4 x 4 1/4"
(7.3 x 14.6 x 10.8 cm)
Manufacturer: Logitech, Inc.
Gift of the manufacturer, 1998

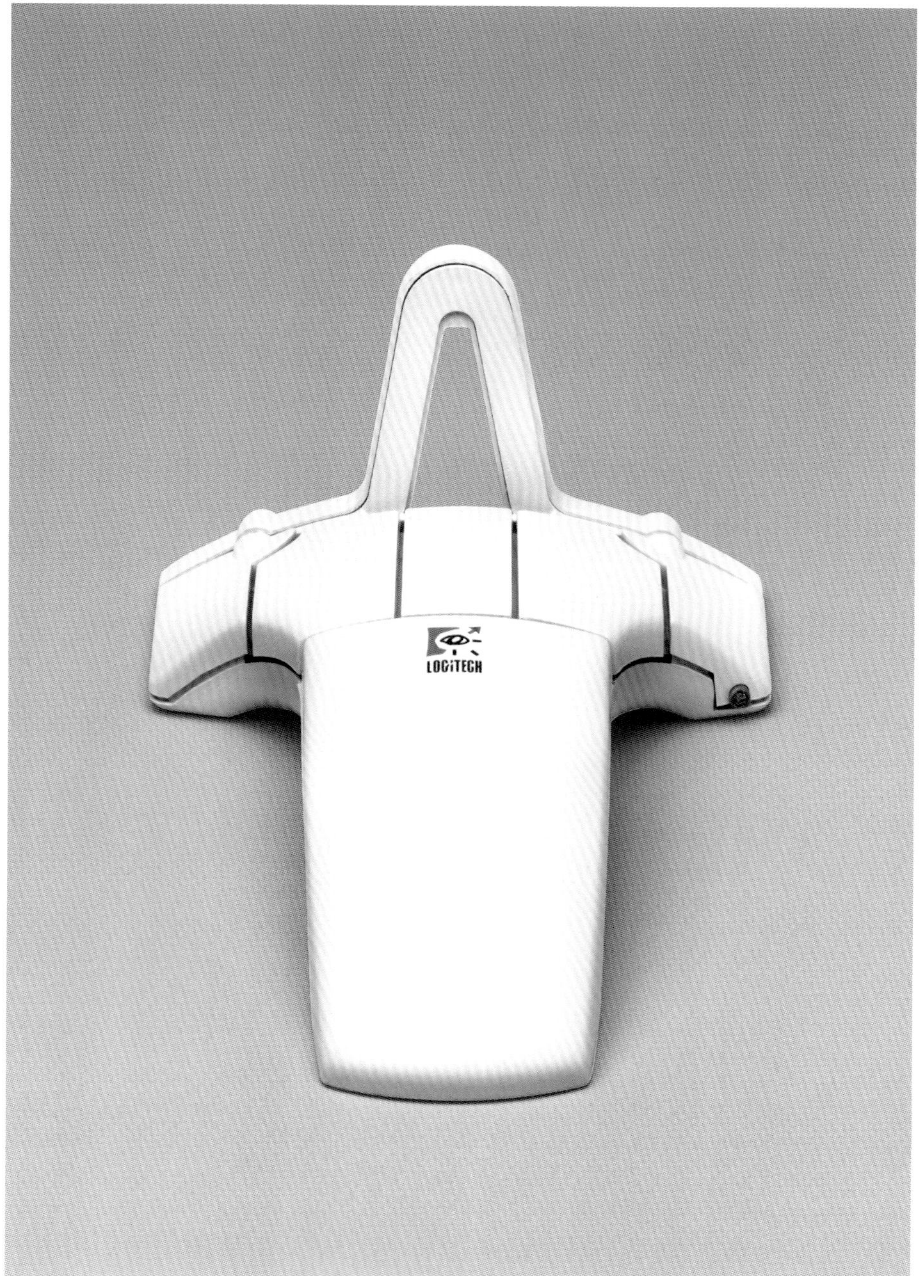

Gordon Randall Perry
Born 1943
ClearVision II Hand-Held Magnifier, 1994 Die-cast urethane and glass, 1½ x 2¼ x 6″ (3.8 x 5.7 x 15.2 cm) Manufacturer: Designs for Vision, Inc., USA Gift of the designer, 1998

Bob Evans

Born 1950

Tan Delta Force Fin Diving Fin, 1994
Polyurethane resin, 17 x 11¼ x 4¼″
(43.2 x 28.6 x 10.8 cm)
Manufacturer: Bob Evans Designs, Inc.
Gift of the manufacturer, 1998

Gregg Fleishman

Born 1947

Chair, 1992
Birch, 43½ x 24⅛ x 23¾″
(110.5 x 61.3 x 60.3 cm)
Purchase, 1993

Steven T. Kaneko

Born 1962

Mouse Computer Pointing Device, 1992
ABS polymer casing,
1½ x 2½ x 4½" (3.8 x 6.3 x 11.4 cm)
Manufacturer: Microsoft Corp.
Gift of the manufacturer, 1993

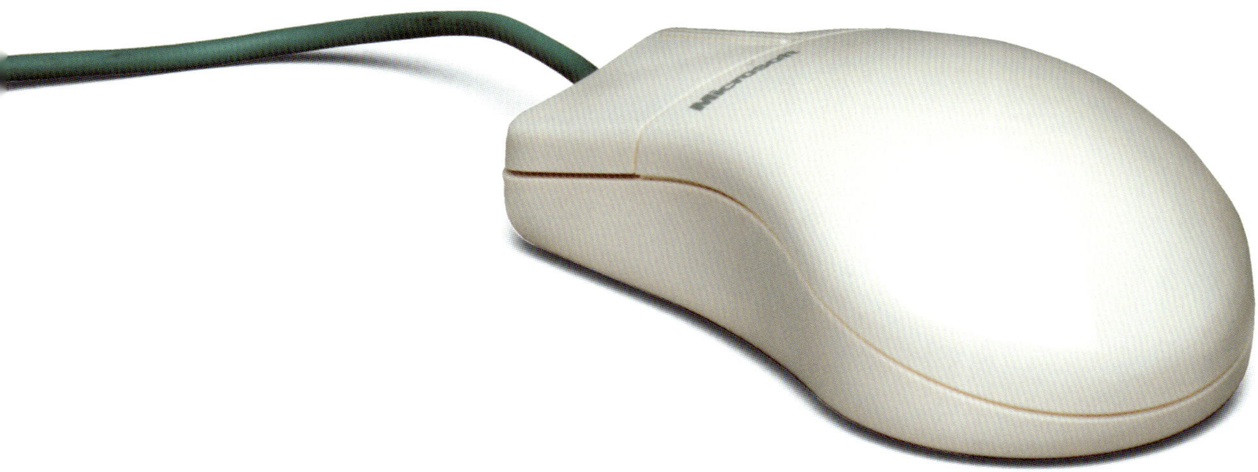

Greg Marting

Born 1960

Giro Atmos Bicycle Helmet, 2003
Polystyrene, polycarbonate,
carbon composite, and nylon,
11 x 8¼ x 6⅜" (28 x 21 x 16 cm)
Manufacturer: Giro Sport Design, Inc.,
a division of Bell Sports, Inc.
Gift of the manufacturer, 2006

Deborah Adler

Born 1975

Klaus Rosburg

German, born 1962

Target ClearRx Prescription System, 2004
Polyethylene terephthalate and paper, 2¼ x 4 x 1½″ (5.8 x 10.2 x 3.8 cm)
Manufacturer: Setco, Inc., a division of Kerr Group, Inc.
Gift of Target Corporation, 2006

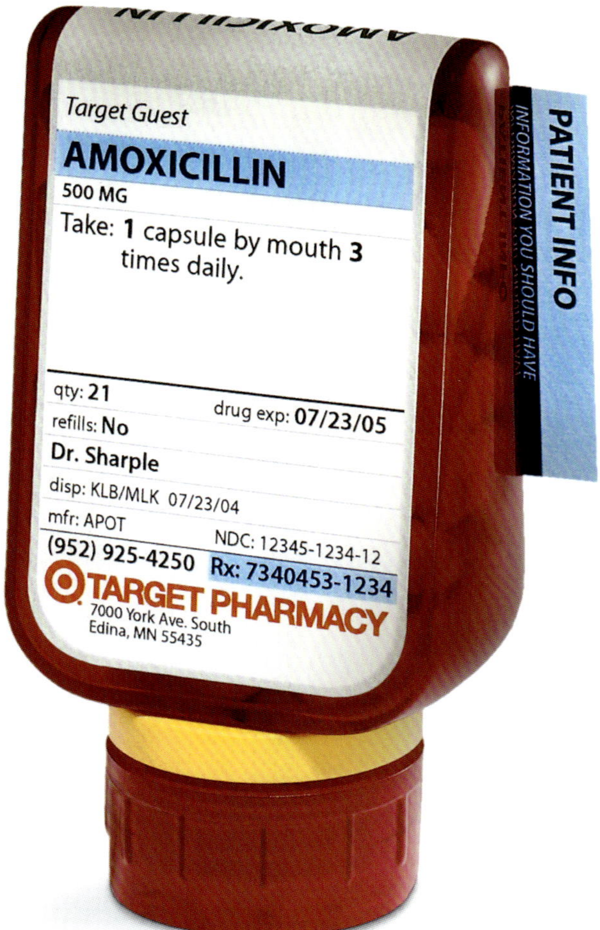

Benjamin Rivera

Born 1967

Wave Multi-Tool, 2004
Stainless steel, closed: l. 4″ (10.1 cm); deployed: l. 6¼″ (15.8 cm)
Manufacturer: Leatherman Tool Group, Inc.
Gift of the manufacturer, 2006

Timeline

1880–1889

Oscar Wilde launches his tour of the United States (1881)

Thomas Edison begins construction of the Pearl Street power station, New York (1882)

The Supreme Court finds the Civil Rights Act of 1875 unconstitutional; construction of the Brooklyn Bridge is completed (1883)

Westinghouse builds alternating-current power plant in Buffalo using technology developed by Nikola Tesla (1886)

Emile Berliner patents the disc phonograph (1887)

The Eiffel Tower is completed and is the landmark of the Exposition Universelle, Paris (1889)

1890–1899

Battle of Wounded Knee, South Dakota (1890)

Louis Sullivan's Transportation Building appears at the World's Columbian Exposition ("The White City"), Chicago (1893)

The Supreme Court upholds racial segregation under the doctrine of "separate but equal" (1896)

The Spanish-American War begins, following the explosion of the battleship *USS Maine* in Havana Harbor, Cuba (1898)

Showman W. G. Crush collides two six-car trains of obsolete rolling stock near West, Texas. The locomotives' boilers explode, killing three and injuring others in the crowd of over 30,000 observers (1899)

1900–1909

The Exposition Universelle, Paris, is the high-water mark for Art Nouveau in Europe; the Kodak Brownie camera debuts, retailing for one dollar (1900)

President William McKinley is assassinated (1901)

Wright brothers' first successful powered flight (1903)

A massive earthquake and fires devastate San Francisco; Frank Lloyd Wright's Unity Temple is completed in Chicago (1906)

Leo Baekeland perfects Bakelite, the first permanent thermosetting plastic (1907)

The National Association for the Advancement of Colored People (NAACP) is founded (1909)

1910–1919

Three-way race for president among Theodore Roosevelt (Progressive), William Howard Taft (Republican), and Woodrow Wilson (Democrat) (1912)

First mass-produced Model T Ford (1913)

World War I begins (1914)

The United States enters World War I (1917)

World War I ends (1918); worldwide influenza epidemic kills 50 to 100 million people over the following two years

Prohibition begins; fear of a Bolshevik revolution in the United States creates the first Red Scare (1919)

1920–1929

Warren G. Harding is elected president, promising "A return to normalcy" (1920)

Harding dies and is succeeded by Calvin Coolidge (1923); an estimated one million listeners tune in to hear him speak on the radio (1924)

Exposition Internationale des Arts Décoratifs et Industriels Moderne, Paris, the "Art Deco" International Exhibition (1925)

Charles Lindbergh flies solo across the Atlantic Ocean; Harley Earl designs the LaSalle convertible body (1927)

Norman Bel Geddes and Walter Dorwin Teague establish industrial design offices in New York, founding the profession in the United States (1927)

Wall Street financial collapse ushers in the Great Depression; The Museum of Modern Art opens in New York (1929)

1930–1939

The depths of the Great Depression; *Modern Architecture: International Exhibition* opens at MoMA (1932)

Franklin D. Roosevelt is inaugurated as president; his "100 days" legislation restores public confidence; Prohibition is repealed (1933)

Machine Art opens at MoMA; American designers organize the National Alliance of Art and Industry exhibition in New York (1934)

The *Hindenburg* airship burns in Lakehurst, New Jersey, killing thirty-six people (1937)

The New York World's Fair ("The World of Tomorrow") opens; World War II begins; new MoMA building opens at 11 West 53rd Street in New York (1939)

1940–1949

The United States enters World War II following a Japanese attack on Pearl Harbor, Hawaii (1941)

Allied forces invade Normandy, France; the Society of Industrial Designers is established (1944)

Germany is defeated; Japan surrenders following the atomic bombings of Hiroshima and Nagasaki; the United Nations is established (1945)

The long-playing record (LP) is introduced; the National Security Act establishes the Central Intelligence Agency (CIA) (1947)

President Harry Truman desegregates the United States Armed Forces (1948)

Marcel Breuer's House in the Museum Garden opens at MoMA (1949)

1950–1959

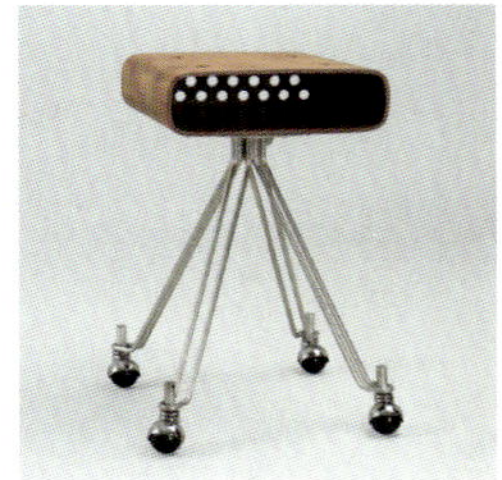

The Korean War begins; the first MoMA *Good Design* exhibition opens in New York and Chicago (1950)

The exhibition *Eight Automobiles* opens at MoMA (1951); *Ten Automobiles* follows in 1953

Julius and Ethel Rosenberg are executed as Soviet spies after being convicted of espionage (1953)

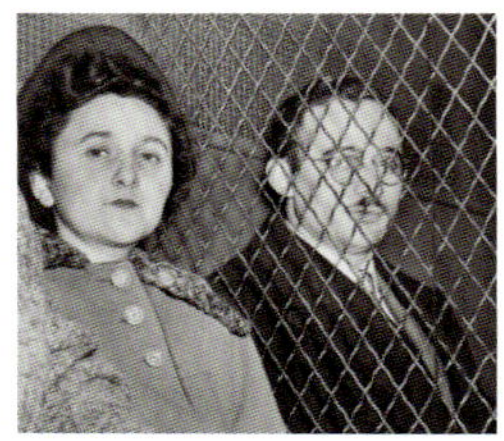

Brown v. Board of Education of Topeka ruling desegregates American schools; the first issue of *Industrial Design* magazine is published (1954)

The USSR launches the *Sputnik* satellite (1957)

Pan American Airways inaugurates trans-Atlantic jet service (1958)

1960–1969

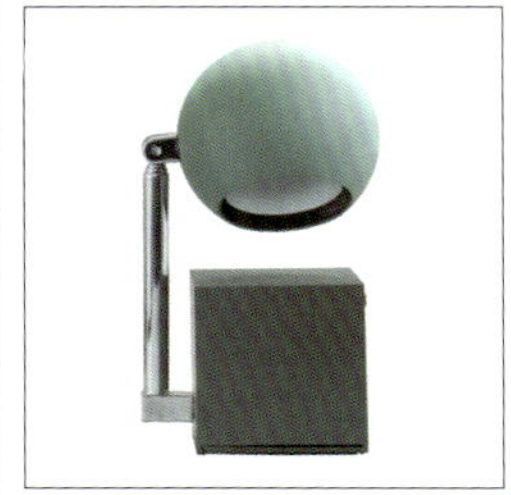

John F. Kennedy is inaugurated as president; Cuban exiles attempt to invade Cuba at the Bay of Pigs; JFK delivers "Ich bin ein Berliner" speech in West Berlin (1961)

JFK is assassinated in Dallas, Texas (1963)

Lyndon Johnson is elected president in Democratic landslide; the New York World's Fair ("Peace Through Understanding") opens (1964)

The US Army quells a race riot in Los Angeles; Malcolm X is assassinated (1965)

High-water mark of Pop art in the United States; *Toward a Rational Automobile* opens at MoMA (1966)

Robert Kennedy and Martin Luther King, Jr., are assassinated (1968)

Woodstock Music and Art Fair, in Bethel, New York; American astronauts walk on the moon (1969)

1970–1979

Operation Linebacker II (the "Christmas Bombings") over North Vietnam is the largest bombing strike by United States forces since World War II; President Richard Nixon visits China and the USSR (1972)

The World Trade Center is completed in New York (1973)

Nixon resigns the presidency; his successor, Gerald Ford, promises that "our long national nightmare is over" (1974)

The last American soldiers leave Saigon on April 30, ending US involvement in Vietnam; *The Architecture of the École des Beaux-Arts* opens at MoMA (1975)

New Wave follows on the heels of punk in the United States and United Kingdom (1978)

1980–1989

Ronald Reagan is elected president; the United States boycotts the Moscow Olympic Games (1980)

The first personal computer, by IBM, comes to market; assassination attempts are made on Reagan and Pope John Paul II (1981)

Reagan proposes "Star Wars" missile defense system (1983)

The space shuttle *Challenger* explodes shortly after takeoff; *Vienna 1900: Art, Architecture and Design* opens at MoMA (1986)

The Dow Jones average plunges 508 points on October 19, the worst one-day loss in the history of the New York Stock Exchange (1987)

Rush Limbaugh's national radio show debuts on WABC, New York (1988)

The Berlin Wall is demolished; China quashes uprising at Tiananmen Square (1989)

1990–1999

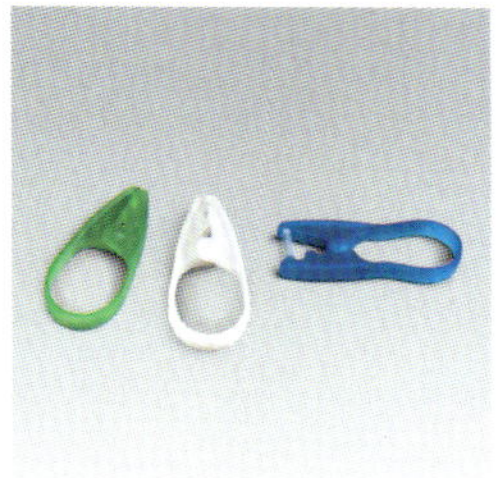

The Hubble Space Telescope is launched into orbit (1990)

The First Gulf War ends 100 hours after Coalition ground forces enter Iraq; the USSR is disbanded; the first Web browser is developed; Nirvana's *Nevermind* introduces alternative rock to a national audience (1991)

Riots follow the verdict in the Rodney King police-brutality trial in Los Angeles; Bill Clinton is elected president (1992)

The Alfred P. Murrah Federal Building in Oklahoma City is bombed, killing 168; *Mutant Materials in Contemporary Design* opens at MoMA (1995)

2000–2007

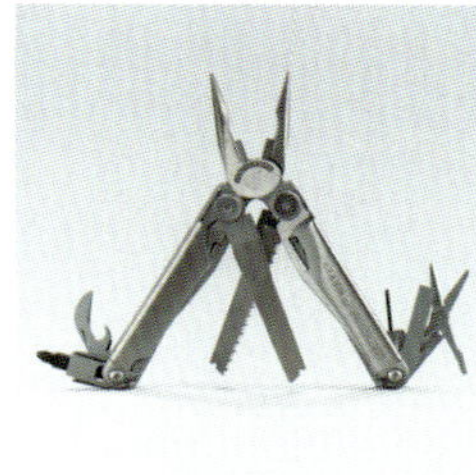

A five–four Supreme Court decision ends efforts to recount election ballots in the state of Florida; George W. Bush is declared president (2000)

On September 11, attacks on the World Trade Center in New York destroy both towers; the Apple iPod MP3 player is launched (2001)

NASA's Mars Odyssey probe begins mapping the surface of the red planet (2002)

Christo and Jeanne-Claude's artwork *The Gates* in Central Park marks New York's post–September 11 resurgence (2005)

Bibliography

ANTONELLI, PAOLA. *Humble Masterpieces.* New York: ReganBooks, 2005.

ANTONELLI, PAOLA. *Mutant Materials in Contemporary Design.* New York: MoMA, 1995. Exh. cat.

ANTONELLI, PAOLA, et al. *Objects of Design from The Museum of Modern Art.* New York: MoMA, 2003.

BANHAM, REYNER, and PENNY SPARKE. *Design by Choice.* New York: Rizzoli, 1981.

BENTON, TIMOTHY. "The Myth of Function." In *Modernism in Design,* edited by Paul Greenhalgh. London: Reaktion Press, 1990.

BRUCE, GORDON. *Eliot Noyes: A Pioneer of Design and Architecture in the Age of American Modernism.* New York: Phaidon, 2006.

CLARK, ROBERT JUDSON. *Design in America: The Cranbrook Vision, 1925–1950.* New York: Harry N. Abrams, 1983.

DOBLIN, JAY. *One Hundred Great Product Designs.* New York: Van Nostrand Reinhold Co., 1970.

DORMER, PETER. *Design Since 1945.* New York: Thames & Hudson, 1993.

FLINCHUM, RUSSELL A. *Henry Dreyfuss, Industrial Designer: The Man in the Brown Suit.* New York: Rizzoli/National Design Museum, 1997. Exh. cat.

FLINK, JAMES J. *The Automobile Age.* Cambridge, Mass.: MIT Press, 1988.

FORTY, ADRIAN. *Objects of Desire: Design & Society from Wedgwood to IBM.* New York: Pantheon Books, 1986.

GORDON, ROBERT B., and PATRICK M. MALONE. *The Texture of Industry: An Archeological View of the Industrialization of North America.* New York: Oxford University Press, 1994.

HANKS, DAVID, and ANNE HOY. *American Streamlined Design: The World of Tomorrow.* Paris: Flammarion, 2005.

HIESINGER, KATHRYN B., and GEORGE H. MARCUS. *Landmarks of Twentieth-Century Design: An Illustrated Handbook.* New York: Abbeville Press, 1993.

HOKE, DONALD R. *Ingenious Yankees: The Rise of the American System of Manufactures in the Private Sector.* New York: Columbia University Press, 1990.

HOUNSHELL, DAVID A. *From the American System to Mass Production, 1800–1932: The Development of Manufacturing Technology in the United States.* Baltimore, Md.: Johns Hopkins University Press, 1984.

KAPLAN, WENDY, ed. *Designing Modernity: The Arts of Reform and Persuasion, 1885–1945.* New York: Thames & Hudson/ The Wolfsonian, 1995.

KIRKHAM, PAT. *Charles and Ray Eames: Designers of the Twentieth Century.* Cambridge, Mass.: MIT Press, 1995.

LAMM, MICHAEL, and DAVE HOLLS. *A Century of Automotive Style.* Stockton, Calif.: Lamm-Morada Publishing Co., 1996.

LICHTENSTEIN, CLAUDE, and FRANZ ENGLER. *Streamlined: A Metaphor for Progress.* Baden, Switzerland: Lars Müller Publishers, 1994.

LUPTON, ELLEN, and J. ABBOTT MILLER. *The Kitchen, the Bathroom, and the Aesthetics of Waste: A Process of Elimination.* Cambridge, Mass.: MIT List Visual Arts Center, 1992.

MARCHAND, ROLAND. "The Designers Go to The Fair: Walter Dorwin Teague and the Professionalization of Corporate Industrial Exhibits, 1933–1940." *Design Issues* 8, no. 1 (1991): 4–17.

MARCUS, GEORGE H. *Design in the Fifties: When Everyone Went Modern.* New York: Prestel, 1998.

MARCUS, GEORGE H. *Functionalist Design: An Ongoing History.* New York: Prestel, 1995.

MEIKLE, JEFFREY L. *Design in the USA.* New York: Oxford University Press, 2005.

MEIKLE, JEFFREY L. *Twentieth Century Limited: The Industrial Designer in America, 1925–1939.* Philadelphia: Temple University Press, 1979.

PETROSKI, HENRY. *The Pencil: A Study in Design and Circumstance.* New York: Alfred A. Knopf, 1990.

PULOS, ARTHUR. *The American Design Adventure: 1940 to 1975.* Cambridge, Mass.: MIT Press, 1983.

PULOS, ARTHUR. *American Design Ethic: A History of Industrial Design to 1940.* Cambridge, Mass.: MIT Press, 1983.

PUTNAM, TIM. "The Theory of Machine Design in the Second Industrial Age." *Journal of Design History* I, no. 1 (1988): 25–34.

SEXTON, RICHARD. *American Style: Classic Product Design from Airstream to Zippo.* San Francisco: Chronicle Books, 1987.

WALLANCE, DON. *Shaping America's Products.* New York: Reinhold Publishing Corp., 1956.

WOODHAM, JONATHAN M. *Twentieth-Century Design.* New York: Oxford University Press, 1997.

Biographies

Russell Flinchum

Russell Flinchum, a native of Winston-Salem, North Carolina, received degrees in English literature and art history from the University of North Carolina prior to moving to New York, where he studied at the Graduate Center of the City University of New York. He received a PhD in art history in 1998. Flinchum organized the exhibition *Henry Dreyfuss: Directing Design* at the Cooper-Hewitt National Design Museum, New York, and is the author of *Henry Dreyfuss, Industrial Designer: The Man in the Brown Suit,* both 1997. He teaches in the MFA in Design Criticism program at the School of Visual Arts, New York, and lectures for the Department of Education at The Museum of Modern Art. He has been the archivist of the Century Association Archives Foundation since 1999.

Paola Antonelli

Paola Antonelli is Senior Curator in the Department of Architecture and Design at The Museum of Modern Art, where she has worked since 1994. Her first exhibition for the museum, *Mutant Materials in Contemporary Design* (1995), was followed by many successful shows, including *Thresholds: Contemporary Design from the Netherlands* (1996), *Achille Castiglioni: Design!* (1997), *Humble Masterpieces* (2004), *Safe: Design Takes on Risk* (2005), and *Design and the Elastic Mind* (2008). Antonelli has taught at the University of California, Los Angeles, and at Harvard's Graduate School of Design, and has lectured extensively on design throughout the world.